T0178469

SpringerBriefs in Materials

The SpringerBriefs Series in Materials presents highly relevant, concise monographs on a wide range of topics covering fundamental advances and new applications in the field. Areas of interest include topical information on innovative, structural and functional materials and composites as well as fundamental principles, physical properties, materials theory and design. SpringerBriefs present succinct summaries of cutting-edge research and practical applications across a wide spectrum of fields. Featuring compact volumes of 50 to 125 pages, the series covers a range of content from professional to academic. Typical topics might include

- A timely report of state-of-the art analytical techniques
- A bridge between new research results, as published in journal articles, and a contextual literature review
- A snapshot of a hot or emerging topic
- An in-depth case study or clinical example
- A presentation of core concepts that students must understand in order to make independent contributions

Briefs are characterized by fast, global electronic dissemination, standard publishing contracts, standardized manuscript preparation and formatting guidelines, and expedited production schedules.

More information about this series at http://www.springer.com/series/10111

Aneeya Kumar Samantara • Satyajit Ratha

Metal Oxides/Chalcogenides and Composites

Emerging Materials for Electrochemical Water Splitting

Aneeya Kumar Samantara
School of Chemical Sciences
National Institute of Science Education
and Research
Khordha, Odisha, India

Satyajit Ratha
School of Basic Sciences
Indian Institute of Technology
Bhubaneswar, Odisha, India

ISSN 2192-1091   ISSN 2192-1105   (electronic)
SpringerBriefs in Materials
ISBN 978-3-030-24860-4   ISBN 978-3-030-24861-1   (eBook)
https://doi.org/10.1007/978-3-030-24861-1

This Springer imprint is published by the registered company Springer Nature Switzerland AG
The registered company address is: Gewerbestrasse 11, 6330 Cham, Switzerland

Dr. Aneeya Kumar Samantara would like to dedicate this work to
his parents, Mr. Braja Bandhu Dash and Mrs. Swarna Chandrika Dash, and his beloved wife, Elina.

Dr. Satyajit Ratha would like to dedicate this work to
his parents, Mrs. Prabhati Ratha and Mr. Sanjaya Kumar Ratha.

Preface

With its high specific energy, zero emission, and lightweight nature, hydrogen can lead the green energy revolution. Internal combustion engines running on hydrogen fuel cells have at least twofold higher specific energy than those running on gasoline. Therefore, hydrogen storage is highly essential for next-generation transportation fuel, considering the depleting fossil fuel reserves (and adverse effects arising from their combustion), thus escalating its industrial necessity. However, the limited reserve of hydrogen in the atmosphere has long been a limiting factor to realize hydrogen energy at large scale. In this context, splitting of water through electrochemical processes or solar power-driven techniques is paving the way for next-generation hydrogen production. Similar approach has also been taken toward the production of oxygen which helps in the combustion of hydrogen in hydrogen fuel cells and is also vital for the metal-air battery systems. State-of-the-art catalysts for hydrogen evolution reaction (Pt/C), and oxygen evolution reaction (IrO_2), make use of precious metals and thus are not suitable for industrial-scale production of both hydrogen and oxygen. However, several reports have tried to address the typical challenge to find the low-cost alternatives for these precious metal-based catalysts. Large-scale water splitting is essential not only for sustainable growth, promoting green energy, but also to reduce both carbon footprint and greenhouse gases. Photoelectrochemical water splitting is an effective and promising method to produce hydrogen and oxygen from water. A wide variety of metal-based compounds, especially compounds having core cations from the transition (or d-block) group, have been investigated for their feasibility to be used as efficient catalysts for water splitting reactions. These include carbon-based nanocomposites (metal-free catalysts), metal chalcogenides, oxides, phosphides, borides, carbides, nitrides, hydroxides, and so forth. Transition metal chalcogenides, as low-cost alternatives, have performed well as hydrogen evolution reaction (HER) catalysts. Their stability issues, however, still possess stiff challenges for the batch production process of hydrogen. In this context, metal oxide-based compounds have been investigated over the years along with the metal chalcogenides, and these oxide materials have been found to possess excellent stability even for over thousands of reaction cycles. Several other materials that have been mentioned, e.g., metal phosphides, borides, carbides, etc., have also been implemented as electrocatalysts for HER.

This brief provides a detailed emphasis on the fundamentals of the water splitting process, underlying mechanism, and a niche of catalyst materials/composites to showcase their effectiveness toward both cost optimization and stability. Also, a rigorous comparison has been drawn to show the versatile nature of these catalytic compounds that can provide sufficient flexibility to facilitate the hydrogen production process.

Khordha, Odisha, India   Aneeya Kumar Samantara
Bhubaneswar, Odisha, India   Satyajit Ratha

About the Book

Hydrogen as fuel has a significant role to play in sustainable growth and as energy carrier can decarbonize the energy sector, both at household and industrial level. The current hydrogen economy is not promising, considering few strategic bottlenecks such as the methods used for hydrogen production, storage, and transportation. Also, the move from gray hydrogen to green hydrogen is not going to be an easy task as these projects are still at a nascent stage (though laboratory-scale productions have been technically proven). Thus, to see hydrogen as a clean and carbon-free source of fuel and as an industrial commodity, focus should be shifted toward the hydrogen generation from water through electrolysis technique. Though can be costly at initial stages, electrolysis through electrolyzers can solve both small- and large-scale energy requirements.

This book provides a detailed emphasis on the role of electrolysis for hydrogen production and materials that could prove significant to catalyze the electrolysis process. Various metal compounds such as metal oxides, sulfides, carbides, and phosphides and few carbon-based nonmetallic compounds have been scrutinized for their effectiveness in promoting the hydrogen evolution during water splitting process. These low-cost metallic/carbonaceous materials provide a wide range of alternatives to the state-of-the-art catalyzers based on precious metal components such as Pt and Ir, which could make the concept of hydrogen, being a sustainable fuel and energy carrier, realistic. This is critical for sustainable growth by limiting the consumption of fossil fuels, curbing greenhouse gases like CO_2, and achieving 100% decarbonized energy generation.

Contents

About the Authors

Aneeya Kumar Samantara is presently working as a postdoctorate fellow at the *School of Chemical Sciences, National Institute of Science Education and Research*, Khordha, Odisha, India. He pursued his PhD at *CSIR-Institute of Minerals and Materials Technology*, Bhubaneswar, Odisha, India. Before joining the PhD, he completed the Master of Philosophy (MPhil) in Chemistry from *Utkal University* and Master in Science in *Advanced Organic Chemistry* at *Ravenshaw University*, Cuttack, Odisha. His research interest includes the synthesis of metal oxide/chalcogenides and graphene composites for energy storage and conversion application. To his credit, he has authored 21 peer-reviewed international journals, 6 books (2 books in Springer, 2 in Arcler Press, and 2 in Global Publishing House, India), and 4 book chapters. Four books and two research journals are in press and will probably be published soon.

Satyajit Ratha pursued his PhD at the School of Basic Sciences, Indian Institute of Technology Bhubaneswar, India. Prior to joining IIT Bhubaneswar, he received his Bachelor of Science, First Class Honors, from Utkal University in 2008 and Master of Science from Ravenshaw University in 2010. His research interests include two-dimensional semiconductors, nanostructure synthesis, applications, energy storage devices, and supercapacitors. He has authored and coauthored about 20 peer-reviewed international journals and 4 books (2 in Springer and 2 in Arcler Press).

Abbreviations

1D	One-dimensional
2D	Two-dimensional
3D	Three-dimensional
BET	Brunauer-Emmett-Teller
BMW	Bayerische Motoren Werke
BSCF	$Ba_{0.5}Sr_{0.5}Co_{0.8}Fe_{0.2}O_{3-\delta}$
BTMPs	Binary metal phosphide
CAES	Compressed air energy storage
CCS	Carbon capture and storage
C_{dl}	Double-layer capacitance
CE	Counter electrode
CFP	Carbon fiber paper
CNT	Carbon nanotube
CPE	Constant phase element
CV	Cyclic voltammogram
CVD	Chemical vapor deposition
DFT	Density functional theory
DHE	Dynamic hydrogen electrode
DPs	Double perovskites
ECSA	Electrochemical active surface area
EIS	Electrochemical impedance spectroscopy
EXAFS	Extended X-ray absorption fine structure
FCEV	Fuel cell electric vehicles
FE	Faradic efficiency
FFT	Fast Fourier transforms
HAADF	High-angle annular dark-field
HER	Hydrogen evolution reaction
HRTEM	High-resolution transmission electron microscope
LDH	Layered double hydroxide
LSV	Linear sweep voltammogram
MOF	Metal organic framework

NCO	$NiCo_2O_4$
NHE	Normal hydrogen electrode
NPCC	N, P-codoped carbon shell
NREL	National Renewable Energy Laboratory
NRs	Nano rods
NSs	Nano sheets
NWs	Nano wires
OCP	Open circuit potential
OER	Oxygen evolution reaction
PCN	Phosphorus-doped graphitic carbon nitride
PEM	Polymer electrolyte membrane
PVD	Physical vapor deposition
RDS	Rate determining step
RE	Reference electrode
RG	Reduced graphene oxide
RHE	Reversible hydrogen electrode
RRDE	Rotating ring disk electrode
SCE	Saturated calomel electrode
SEM	Scanning electron microscope
S_{geo}	Geometrical surface area
$S_{geo.}$	Geometrical surface area
SMR	Steam methane reforming
STEM	Scanning tunneling electron microscope
TMC	Transition metal chalcogenide
TMN	Transition metal nitride
TMO	Transition metal oxide
TMP	Transition metal phosphide
TMPS	Transition metal phosphosulfide
TOF	Turnover frequency
UPS	Ultraviolet photoemission spectroscopy
WE	Working electrode
XAS	X-ray absorption spectroscopy
XPS	X-ray photoelectron spectroscopy
XRD	X-ray diffraction

Abstract This book presents a brief discussion on issues of traditional energy resources and types of nonrenewable/alternative energy resources. As an efficient device, more focus has been given to electrolyzers. Further fundamental discussion on different parameters used to evaluate performance of an electrocatalyst is elaborately presented, followed by emergence of different electrocatalysts. Although the noble metal-based catalysts perform efficiently, their high cost and limited reserve motivate the researchers to think some alternatives. In this regard, low-cost transition metal-based electrocatalysts and composites with different conductive carbon materials were developed, but their catalytic activity yet remained far from that shown by the noble metals. Therefore, discussion on the synthesis, mechanism study, and catalytic performance of different electrocatalysts suitable for electrolyzer has been emphasized. The authors presume that this book will help the energy and materials researchers to gather more knowledge in this field and to explore new electrocatalysts with higher efficiencies for commercial application.

Keywords Electrolyzer · Hydrogen evolution · Oxygen evolution reaction · Energy conversion · Metal oxides · Metal chalcogenides

Chapter 1
Introduction

Abstract This chapter presents a brief discussion of different types of energy resources and emergence of electrolysis. Both the advantages and disadvantages of traditional fossil fuel are presented along with the benefits of water electrolyzer. Further the types of electrolysis and requirements to improve their performance has been discussed. We presume that the authors will get fundamental idea on energy issues and requirement of alternative energy resources for future sustainability.

Keywords Fuel cell · Electrolysis · Solar cell · Photovoltaic · Fossil fuel

Energy and environment are two key issues associated with sustainable development of society. Till now fossil fuels (natural gases, oil and coal etc.) are assumed as the main source of energy and it covers around 85% of global energy demands. Though these are valuable natural sources for many of the chemical industries, but their limited reserve demands to think about renewable energy resources. Further, the combustion of fossil fuels emits greenhouse gasses (like CO_2, SO_x, NO_x etc.) causing serious environmental problems. These issues triggered the interest of the energy researchers to develop an alternative, durable, ecofriendly fuel combustion technology with zero carbon emission renewable energy resource. In this regard, solar and wind energy were harvested to generate electricity for the fulfillment of day today energy demands. But seasonal operation and lower efficiency restricted these technologies from continuous energy generation. Thus these technologies are alleviated from list of primary energy sources. Moreover due to the diffusive nature of sunlight, efficient energy capture and storage units are strictly needed for wide application. Therefore photovoltaic devices were developed that converts expeditiously the solar energy to electricity. But the storage and transportation needs highly expensive and bulky batteries. After numerous research efforts, the researchers have successfully stored solar energy in form of chemical energy, especially in form of molecular hydrogen (H_2). Like oil and natural gases, hydrogen is not energy but stores and carries energy. As an efficient energy carrier, H_2 have various advantages like facile storage/transportation, higher specific energy (140 MJ/kg at 700 atm) and combustion takes place without emission of carbon (forms water as combustion product) (Bockris 2002; Conte et al. 2009). In addition to photovoltaic, many processes like stream methane reforming, natural gas oxidation, catalytic (photo chemical/electro chemical) water splitting, biomass electrolysis, carbonation etc. are followed for the production of hydrogen. The steam methane reforming and natural gas oxidation technologies produce around 95% of total hydrogen, but again associated with the use of restricted reserve fossil fuels. And also the emission of CO_2 cannot be avoided if we continue to use these technologies. On the other hand, though the water electrolysis (conversion of water to fuel gas) produce only 4% of hydrogen, but it can be considered as a cleanest way for scalable production of hydrogen.

This conversion of water to fuel gases can be executed either in presence (i.e. photoelectrochemical water electrolysis) or in absence of light (i.e. electrochemical water electrolysis). Further the electrochemical water electrolysis dominant over the photochemical/photoelectrochemical methods. Although the production of hydrogen is the principal objective of water electrolysis, but we cannot ignore oxygen evolution reaction taking place at the counter electrode surface. Therefore on passage of electrical energy, splitting of water molecule to H_2 and O_2 gases takes place. It occurs only when a certain potential reached called as the thermodynamic potential. As per the literature survey, the thermodynamic potential for generation of hydrogen and oxygen are 0 V and ~1.23 V (vs. RHE) respectively. Further the multi proton coupled electron transfer process of OER make the electrolysis to face sluggish reaction kinetics thereby lowering the Faradic efficiency of HER. Thus the electrolysis of water faces very high activation energy barrier requiring additional potential (called overpotential) for their completion. Therefore, main focus of energy researchers lays to reduce this overpotential by improving electrodes, electrocatalysts, electrolytes or by gathering knowledge regarding the kinetics of electrode reactions (Bates et al. 2015; de Souza et al. 2007; Vilekar et al. 2010).

Initially, the noble metals like platinum (Pt), palladium (Pd), ruthenium (Ru) and iridium (Ir) etc. were considered as the only available active electrocatalysts for HER and OER. But their high cost and limited reserve restricts the large scale catalyst production, commercialization and widespread application. The following figure (Fig. 1.1) shows the abundance of HER active elements in the earth crust. One

Fig. 1.1 Crustal abundance of metals that are used for constructing HER electrocatalysts. (Reproduced with permission from Zou and Zhang 2015)

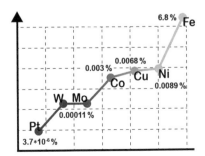

can correlate their abundance with the cost of electrocatalysts and should attempt to design low cost electrocatalyst material for commercial production of hydrogen. Since the existence of Pt is many orders of magnitude smaller ($3.7 \times 10^{-6}\%$) than the non-precious metals, so these electrocatalysts are of very high cost restricting the commercial production. On the other hand, the non-noble metals exists as in the order of W = Mo<Co<Cu<Ni<<Fe. Among them the Ni and Fe have higher abundance and low price. Therefore the synthesis of Ni and Fe based electrocatalysts is recommended for the economical hydrogen production.

In this regard, transition metal based electrocatalysts found to be the suitable alternative materials. Therefore much more effort have been devoted and the transition metal based chalcogenides, oxides, hydroxides, phosphides, nitrides, carbides etc. and their composites with conductive carbon nanomaterials like graphene, carbon nanotubes etc. were developed. Since these HER and OER are the surface reactions, the morphology of electrocatalyst found to play an important role in their catalytic performance. Therefore the quest is not limited to find new electroactive materials but also the best surface morphology. By varying the starting precursors and reaction conditions, nanostructures with various surface morphologies have been successfully developed. Specifically, one dimensional porous nano rods/nano wires, two dimensional nano sheets/flake like structures and three dimensional nano spheres studied exclusively. Some groups also achieved an impressive catalytic activity by incorporating hetero atoms/ions into the lattice of the nanostructures. But their catalytic efficiencies still remain far apart from the noble metal based electrocatalysts. Therefore development of new electroactive material with better catalytic efficacy is strictly indispensible for smooth running of these electrolyzers.

Generally the catalysts are of powdery in nature and the electrodes were prepared either by drop casting or by forming films onto the conductive surface. But during the course of OER/HER, the fast evolving gases leads to peel the catalyst film out from the electrode surface showing a sudden fall in catalytic performance. In order to mitigate these issues, free standing and supported electrodes were developed. They show a strong adherence of the active catalyst on the electrode surface thereby showing excellent long term cyclic stability.

Most of the hydrogen is used in various industries such as for the refining of petroleum, synthesis of urea from ammonia etc. Nowadays hydrogen powered vehicles (fuel cell electric vehicles; FCEV) are developed showing similar performances like combustion engines and offers zero carbon emission. Therefore many of the world's leading car manufacture companies such as BMW, Ford, Audi, Toyota and Hyundai putting their sustainable effort to bring these FCEVs in our realistic future. As an emerging field of energy sector, there are many review articles and perspectives on electrocatalyst synthesis and application are available (Anantharaj et al. 2016; Tahir et al. 2017; Zou and Zhang 2015). Some of them have focused on the synthesis of the active materials where others have given emphasis to compare the electrocatalytic performances. Hence for better understanding on the electrolysis processes, associated reaction mechanisms, materials activity and their commercial application in a single platform is needed. In this view we have prepared this book that covers all the fundamental aspects as well as up to date literature survey on this emerging field for energy researchers. Starting from the types and fundamentals of water electrolysis, a detailed reaction mechanism at different electrolytic conditions is presented in the first section of the book. Further a complete literature survey on recent research and development of precious metal catalysts, transition metal oxides/hydroxides, chalcogenides, phosphides, supported and free standing electrodes etc. is presented to up to date the readers. Then the advancement in scalable production and commercial application of hydrogen fuel in day to day life is presented. The book concludes with a brief summary and scope available to work in this field for sustainable future.

References

Anantharaj, S., Ede, S. R., Sakthikumar, K., Karthick, K., Mishra, S., & Kundu, S. (2016). Recent trends and perspectives in electrochemical water splitting with an emphasis on sulfide, selenide, and phosphide catalysts of Fe, Co, and Ni: A review. *ACS Catalysis, 6*, 8069–8097.

Bates, M. K., Jia, Q., Ramaswamy, N., Allen, R. J., & Mukerjee, S. (2015). Composite Ni/NiO-Cr2O3 catalyst for alkaline hydrogen evolution reaction. *Journal of Physical Chemistry C, 119*, 5467–5477.

Bockris, J. O. (2002). The origin of ideas on a hydrogen economy and its solution to the decay of the environment. *International Journal of Hydrogen Energy, 27*, 731–740.

Conte, M., Di Mario, F., Iacobazzi, A., Mattucci, A., Moreno, A., Ronchetti, M., 2009. Hydrogen as future energy carrier: The ENEA point of view on technology and application prospects. Energies.

de Souza, R. F., Padilha, J. C., Gonçalves, R. S., de Souza, M. O., & Rault-Berthelot, J. (2007). Electrochemical hydrogen production from water electrolysis using ionic liquid as electrolytes: Towards the best device. *Journal of Power Sources, 164*, 792–798.

Tahir, M., Pan, L., Idrees, F., Zhang, X., Wang, L., Zou, J.-J., & Wang, Z. L. (2017). Electrocatalytic oxygen evolution reaction for energy conversion and storage: A comprehensive review. *Nano Energy, 37*, 136–157.

Vilekar, S. A., Fishtik, I., & Datta, R. (2010). Kinetics of the hydrogen electrode reaction. *Journal of the Electrochemical Society, 157*, B1040–B1050.

Zou, X., & Zhang, Y. (2015). Noble metal-free hydrogen evolution catalysts for water splitting. *Chemical Society Reviews, 44*, 5148–5180.

Chapter 2
Types of Electrolysis and Electrochemical Cell

Abstract Electrolysis is the process of breakdown of stable water molecule on passage of current across the electrodes in aqueous electrolyte. Generally at a particular potential, the splitting of water to molecular hydrogen and oxygen takes place at cathode and anode terminals of the electrochemical cell. This chapter presents a brief discussion on the types of electrolysis and electrochemical cell. Further, for better understanding, complete configuration of electrodes in a three electrode electrochemical cell is presented.

Keywords HER · OER · Cathode · Anode · Electrochemical cell · Electrode arrangement · Measurement

2.1 Electrolysis

Electrolysis is an electrochemical process performed by passing current among the electrodes through an aqueous electrolyte solution. During the course of reaction, dissociation of stable water molecule takes place forming hydrogen (H_2) and oxygen (O_2) gasses on the electrode surfaces. The overall reaction of water electrolysis is presented as follows,

$$H_2O \rightleftharpoons H_2 + \frac{1}{2}O_2$$

It comprises two half-cell reactions as (i) hydrogen evolution reaction (HER) where the reduction of water takes place forming molecular hydrogen on the cathode surface and (ii) oxygen evolution reaction (OER) involving the oxidation of water on anode electrode surface forming molecular oxygen. Since both HER and OER depends on the nature of active ions (H^+, OH^-), the reactions are different in different pH of the electrolyte solution. In acidic electrolyte, the active ion is proton and in alkaline solution it is a hydroxide ion. The following table presents the complete equations for OER and HER.

© The Author(s), under exclusive license to Springer Nature Switzerland AG 2019
A. K. Samantara, S. Ratha, *Metal Oxides/Chalcogenides and Composites*,
SpringerBriefs in Materials, https://doi.org/10.1007/978-3-030-24861-1_2

Types of reaction	In acidic electrolyte	In alkaline electrolyte
Hydrogen evolution reaction	$4H^+ + 4e^- \rightleftarrows 2H_2\uparrow$	$4H_2O + 4e^- \rightleftarrows 2H_2\uparrow + 4OH^-$
Oxygen evolution reaction	$2H_2O \rightleftarrows O_2\uparrow + 4H^+ + 4e^-$	$4OH^- \rightleftarrows 4H_2O + 2O_2\uparrow + 4e^-$

The detailed reaction mechanisms are elaborately discussed in the following sections. By using cyclic voltammogram technique, Wang et al. have discussed all the possible reactions occurring during the course of electrolysis (Wang et al. 2009). Here a platinum micro electrode is employed as the working electrode and CVs were recorded both in water as well as in 0.5 M H_2SO_4 electrolyte (Fig. 2.1). The electrolysis process begins with (a) adsorption of oxygen on electrode surface forming the Pt-O film followed by the (b) evolution of gaseous oxygen. On decreasing potential during the cathodic scan, the as formed Pt-O gets reduced by (c) desorbing oxygen (oxygen reduction). On further decreasing potential, (d) adsorption of hydrogen occurs followed by (e) evolution of molecular hydrogen. Then the hydrogen gets oxidized on increasing the sweeping potential completing the scanning cycle.

The conversion of water to oxygen and hydrogen takes place on passage of current when reached at a particular potential called as the thermodynamic potential. The value of which can be derived from the following equation,

$$\Delta G = -\vartheta F E^0$$

Here,

ΔG = Gibb's free energy (J/mol)
ϑ = Number of electrons consumed during the reaction
F = Faraday constant (96485 C/mol)

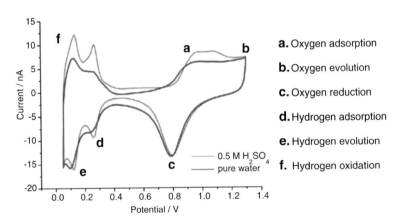

Fig. 2.1 Cyclic voltammogram for pure water and H_2SO_4 on Pt micro-disk electrodes. The scanning rate is 80 mV/s. A DHE/PEM system is used as reference electrode. (This figure is an edited version of a figure from Wang et al. 2009)

Using above equation, the standard potential of electrolysis (combined potential of HER and OER) has been calculated to be 1.229 V. Whereas under the standard conditions, the HER has 0 V and OER has −1.229 V and are combined to give the net potential of the cell as per the following equation,

$$E^0_{cell} = E^0_{cathode} - E^0_{anode}$$

$E^0_{cell} = Standard\, cell\, potential\, (V)$

$E^0_{cathode} = Standard\, potential\, of\, cathodic\, half\, reaction\, (V)$

$E^0_{anode} = Standard\, potential\, of\, anodic\, half\, reaction\, (V)$

Better understanding on the role of reaction condition on electrocatalytic activity, reaction kinetics, and types of active electrocatalysts, activity screening parameters and detailed reaction mechanism can be achieved from the following sections.

2.2   Electrochemical Cell

The catalytic activities of a catalyst are generally studied in a three electrode electrochemical set up as presented in the following figure (Fig. 2.2). The whole set up comprises a catalyst modified working electrode (generally glassy carbon electrode), reference and counter electrode (platinum wire) dipped in a suitable electrolyte of optimized concentration. Sometimes a two compartment (separated by a porous frit) electrochemical cell is preferred and in some cases separate half-cell connected with a salt bridge is used for the water electrolysis study. The preparation of working electrode depends on the type of electrocatalysts we are working on. In case of the powder catalysts, an optimal amount of catalyst mixed with a binder (generally Nafion

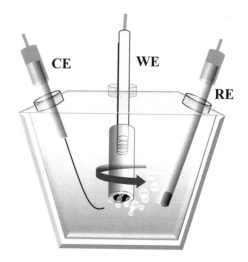

Fig. 2.2 Schematic presentation of a three electrode electrochemical cell used for the electrolysis study. CE counter electrode, WE working electrode, and RE reference electrode

is used) and a solvent to form catalyst ink. To make a uniform dispersion and to increase the conductivity of the catalyst ink, some conductive carbon materials (acetylene black, volcano carbon, graphene etc.) are generally recommended to add to the catalyst ink. Then the ink is drop casted onto the working electrode and measurement can be carried out after completely drying (drying of electrode can be carried out either in room temperature or by using a vacuum oven). But the self-supported catalysts like those prepared by means of electro deposition method or by substrate mediated grown (either by physical or chemical means) can be directly used as working electrode. Though glassy carbon is employed as working electrode in most of the cases, but the titanium plate, nickel foil, nickel foam etc. are used in some reports to adhere the active electrocatalyst for the measurement. Among different reference electrodes, the calomel electrode (in saturated KCl) and silver chloride electrode (in saturated KCl) are commonly used that involves the following redox reactions as,

$$2Hg + Cl \rightleftarrows Hg_2Cl_2$$

$$Ag + Cl \rightleftarrows AgCl$$

But in a single compartment cell, there is a possibility of leaching of chlorine from these reference electrodes and diffuse to the working electrode affecting the catalyst activity. The interference of such chloride ions can be avoided by using the oxide (mercury oxide) or sulphate (mercury sulphate, silver sulphate etc.) based reference electrodes. Further to achieve better performance of the electrocatalyst and to avoid the interference of the ions leaching from the reference electrodes and the gas bubbles evolved from the counter electrode, cells are designed using separating membranes.

Also H-type electrochemical cells are used to make a well separation between the working, reference and counter electrode and to quantify the gasses (H_2 or O_2) evolving during the electrolysis. In these three electrode measurements, the cell potential can be monitored across the working and reference electrode and cell current can be scaled up among the counter and working electrode. In case of the HER/OER, though different reference electrodes are used but all are converted to the reversible hydrogen electrode (RHE) by using the following Nerst equations,

For Ag/AgCl reference,

$$E_{RHE} = E_{Ag/AgCl} + E^0_{Ag/AgCl} + 0.059 \; pH; E^0_{Ag/AgCl} = 0.21V$$

For saturated calomel electrode (SCE),

$$E_{RHE} = E_{SCE} + E^0_{SCE} + 0.059\,pH; E^0_{SCE} = 0.244\,V$$

For mercury oxide electrode,

$$E_{RHE} = E_{Hg/HgO} + E^0_{SCE} + 0.059\,pH; E^0_{Hg/HgO} = 0.897\,V$$

Reference

Wang, Q., Cha, C.-S., Lu, J., & Zhuang, L. (2009). The electrochemistry of "solid/water" interfaces involved in PEM-H2O reactors Part I. The "Pt/water" interfaces. *Physical Chemistry Chemical Physics, 11*, 679–687.

Chapter 3
Mechanism and Key Parameters for Catalyst Evaluation

Abstract Both the hydrogen evolution and oxygen evolution reaction follows a multi electron catalytic path and the mechanism strongly depends on the types of electrolyte used for the electrolysis. Also there are various key parameters available to evaluate the performances of a particular electrocatalyst. In this chapter, detailed discussion on the mechanism of both the HER and OER in acidic and alkaline electrolyte is presented. Moreover, emphasis has been given on the calculation of different key parameters like overpotential, Tafel slope, electrochemical active surface area, Faradic efficiency, Turnover frequency, long cycle life etc. used for efficiency evaluation of a catalyst.

Keywords Acid electrolyte · Alkaline electrolyte · Mechanism · Overpotential · Tafel slope · Faradic efficiency · ECSA · Mass activity · Long cycle life

Both the oxygen and hydrogen evolution reactions follow a multi electron catalytic path in the acid, alkaline and neutral electrolyte medium. Various parameters like overpotential (η), electrochemical active surface area (ECSA), turn-over frequency (TOF), Tafel slope, exchange current density and operational durability etc. are broadly employed to evaluate the catalytic performance of different electrocatalysts. Moreover, these parameters play a vital role to explore the associated mechanism in these electrocatalytic processes. The detailed mechanism and key parameters are discussed in the following sections.

3.1 Mechanism of Oxygen Evolution Reaction

Some of the electrocatalysts show their performances for either of the half-cell reactions (i.e. either OER or HER) or both reactions in acid electrolytes whereas others perform well in alkaline medium. Though various research groups proposed

© The Author(s), under exclusive license to Springer Nature Switzerland AG 2019
A. K. Samantara, S. Ratha, *Metal Oxides/Chalcogenides and Composites*,
SpringerBriefs in Materials, https://doi.org/10.1007/978-3-030-24861-1_3

different reaction mechanisms, but the oxygen evolution reaction follows the four electron reaction path both in the alkaline and acid medium. For a typical catalytic surface (M), the involved mechanisms are as follows,

(i) **In alkaline medium;**

$$M + OH^- \rightarrow MOH \tag{3.1}$$

$$MOH + OH^- \rightarrow MO + H_2O_{(l)} \tag{3.2}$$

$$2MO \rightarrow 2M + O_{2(g)} \tag{3.3}$$

$$MO + OH^- \rightarrow MOOH + e^- \tag{3.4}$$

$$MOOH + OH^- \rightarrow M + O_{2(g)} + H_2O_{(l)} \tag{3.5}$$

And the overall reaction is,

$$4OH^-_{(aq)} \rightarrow 2H_2O_{(aq)} + 4e^- + O_{2(g)}$$

(ii) **In acid medium;**

$$M + H_2O_{(l)} \rightarrow MOH + H^+ + e^- \tag{3.6}$$

$$MOH + OH^- \rightarrow MO + H_2O_{(l)} + e^- \tag{3.7}$$

$$MO \rightarrow 2M + O_{2(g)} \tag{3.8}$$

$$MO + H_2O_{(l)} \rightarrow MOOH + H^+ + e^- \tag{3.9}$$

$$MOOH + H_2O_{(l)} \rightarrow M + O_{2(g)} + H^+ + e^- \tag{3.10}$$

And the overall reaction is,

$$2H_2O_{(aq)} \rightarrow 4H^+_{(aq)} + 4e^- + O_{2(g)}$$

Generally two different routes are followed to generate oxygen from the MO intermediate. In case of acid mediated mechanism, the decomposition of MO (as presented in Eq. 3.8) and the combination of as formed MOOH with H_2O (as presented in Eq. 3.10) leads to generate O_2. But in alkaline electrolyte, the decomposition of MO (Eq. 3.3) and interaction of as formed MOOH with OH⁻ (Eq. 3.5) forms oxygen in an efficient manner. In this heterogeneous catalysis process, all the bonding interactions between M and O within the formed intermediates (MO, MOOH and MOH) are very crucial to determine the overall reaction mechanism. Therefore the binding strength of intermediates to catalyst surface determines the efficiency of that particular material towards OER. But the trouble in experimentally measuring and controlling the binding strength of the adsorbate on catalyst surface make very difficult to predict the catalytic activity of a new catalyst. In this regard substantial efforts have been paid to identify the properties of surface electronic structures which can be used to evaluate the catalytic efficiency and to evolve chemistries of new electrocatalysts (Bockris and Otagawa 1984; Suntivich et al. 2011; Trasatti 1991).

Among other catalytic activity descriptors, the relationship between the strength of chemisorbed oxygen with the metal surface via metal d-band center relative to its Fermi level plays a great role in activity evaluation (Edmonds and McCarroll 1978; Hammer 2006). In this approximation, when the adsorbate binds to the catalyst surface, its electrons interact with the s, p and d-band electrons of the catalyst. Since the s and p-band energies do not change significantly during the metal-adsorbate bond formation, so the localized d-band of the metal only governs the bond strength. As the d-band center shift towards Fermi level, the number of vacant anti-bonding orbitals above the Fermi level increases strengthening the chemical bonding with the adsorbate (Koper and van Santen 1999). According to DFT calculations, a linear relationship among the d-band center and strength of metal (transition metal)-oxygen bonding has been established. Afterwards, this was also supported by the ultraviolet photoemission spectroscopic (UPS) measurements of various alloys of Pt (in form of Pt_3M, M = Ti, V, Cr, Mn, Fe, Co, Ni) (Mun et al. 2005).

The as discussed proton coupled electron transfer reaction pathways are initially proposed for the noble metal surfaces. Thereafter it was extended to justify the catalytic activity trends of the perovskites and rutile oxides towards OER (Fig. 3.1a) (Man et al. 2011). Sometimes these are called as the acid-base mechanism that proceeds via a number of acid-base reaction steps (Betley et al. 2008; Mavros et al. 2014). Here, the oxygen nucleophile OH (Lewis acid) attacks to the electrophilic metal oxide surface (Lewis base) to carry out the oxygen evolution reaction. A similar reaction mechanism has also been proposed for the pyrochlore and rutile oxides (Fig. 3.1b) (Goodenough et al. 1990). The computational calculations suggest this mechanism as a most favorable one for the dimeric metal oxide molecule with early transition metal ions (Fig. 3.1c) (Mavros et al. 2014). During the course of reaction, the metal oxides adsorbs various reactive species like H∗, OH∗, O∗ and etc. from electrolyte depending on the catalyst's surface pH at the zero charge (pH_{pzc}) (Noh and Schwarz 1989; Pechenyuk 1999). When the pH value is higher than pH_{pzc}, then it lead to accumulate the negatively charged species (OH∗, OOH∗, O∗ etc.) onto the oxide surface to govern the OER. It has been observed that the oxides show their activity towards OER in pH of 13–14 that are much higher than their pH_{pzc} values (~7–11 for binary metal oxides and perovxides) (Bockris and Otagawa 1984; Kosmulski 2009). Therefore the catalytic surfaces of these oxides expected to accumulate the OH^- species in the alkaline medium at initial step of the reaction and then proceeds to evolve oxygen (Patrick 2004). However in all the discussed mechanisms, the whole reaction carried out on a single metal site, but latter on a reduced overpotential is observed in two-site mechanism (Halck et al. 2014). Here, reaction

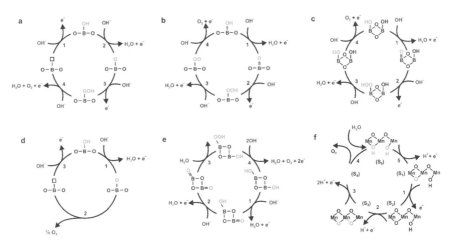

Fig. 3.1 OER mechanisms proposed for crystalline oxide surfaces, amorphous oxide surfaces, molecules and the oxygen evolution complex of photosystem II (OEC-PS II). Except for photosystem II (being in a mildly acidic environment), all catalytic cycles are formulated for alkaline electrolytes. (**a**) Four-step reaction mechanism proposed by Rossmeisl and co-workers for the OER on noble metal catalyst surfaces, (Rossmeisl et al. 2005) and later applied to oxide surfaces (Man et al. 2011). (**b**) Four-step reaction mechanism proposed by Goodenough et al. for the OER on perovskite surfaces (Goodenough et al. 1990). (**c**) Acid–base mechanism proposed for dimeric molecules (Mavros et al. 2014). (**d**) Reaction mechanism proposed by Trasatti and co-workers involving recombination of oxygen atoms to produce O_2 (De Faria et al. 1996). (**e**) Reaction mechanism proposed by Gerken et al. for electrodeposited oxides in buffered conditions (pH 3.5 to 14) (Gerken et al. 2011). (**f**) Structural changes of the OEC-PS II proposed by Dau, Haumann and coworkers based on X-ray absorption spectroscopy (Haumann et al. 2005b). Three of the four manganese ions are sufficient to show the catalytic cycle, which proceeds through states S0 to S4 by light flashes (Haumann et al. 2005a). In all panels, orange denotes species on the catalyst surface and blue denotes species in solution. The transitions are labeled starting from the resting state of the catalyst, i.e. the state in equilibrium with the surroundings in absence of external stimuli (voltage or light). (Reproduced with permission from Hong et al. 2015)

steps like recombination of adsorbate to form O_2 or dissociation of H_2O are not considered as the feasible reaction pathways. This reaction pathway primarily formulated for RuO_2 surfaces in acidic media that involves the recombination of surface oxygen species, later on observed to be competitive in late transition metal based dimeric molecules (i.e. Cu, Ni, Co) (Fig. 3.1d) (De Faria et al. 1996; Mavros et al. 2014). Though it proceed like the acid-base mechanism, but does not involve the potentially rate limiting OOH∗ species. Moreover this reaction path shows a large activation barrier on the noble metal surface compared to the surfaces of metal oxides, electrodeposited oxides and oxygen evolving complexes (the loosely bound oxygen species facilitates the reaction steps) (Fig. 3.1e, f) (Gerken et al. 2011; Haumann et al. 2005a; Hibbert 1980; Nørskov et al. 2002; Wohlfahrt-Mehrens and Heitbaum 1987). The reaction mechanisms thus discussed here are involved the formation of various reaction intermediates, but no such techniques/methods have been evolved to identify the intermediates on the oxide catalytic surfaces.

3.2   Mechanism of Hydrogen Evolution Reaction

The cathodic half reaction i.e. hydrogen evolution reaction takes place either by the reduction of protons (in case of the acidic medium) or water (in alkaline medium) in the electrolyzer during the electrolysis process. The overall reaction path may be either the Volmer-Heyrovsky or Volmer-Tafel as per the following equations,

(i) In alkaline medium;

Since in the alkaline medium the OH^- ions are abundantly present, so the reaction proceeds as per the following steps;

Step-1_*Volmer reaction*: The water molecule combines with an electron forming hydrogen atom that adsorbs to the catalytic active site of the electrode surface.

$$* + H_2O + e^- \rightleftharpoons H_{ad}^* + OH^-$$

Step-2_*Heyrovsky reaction*: The adsorbed hydrogen atom couples with a water molecule and an electron allowing the electrochemical desorption of hydrogen.

$$H_{ad}^* + H_2O + e^- \rightleftharpoons H_2 + OH^- + *$$

Step-3_*Tafel reaction*: In this step two of the adsorbed hydrogen combine with each other liberates one hydrogen molecule as in the following equation.

$$2H_{ad}^* \rightleftharpoons H_2 + 2*$$

(ii) In acid medium;

Step-1_*Volmer reaction:* In this step the H^+ ion combine with an electron and get adsorbed onto the catalytic active site.

$$* + H^+ + e^- \rightleftharpoons H_{ad}^*$$

Step-2_*Heyrovsky reaction:* The above adsorbed hydrogen coupes with an electron and another H+ leading to an electrochemical desorption.

$$H_{ad}^* + H^+ + e^- \rightleftharpoons H_2 + *$$

Step-3_*Tafel reaction:* As in the case of alkaline medium, here also two adsorbed hydrogen get combine with each other and lead to liberate molecular hydrogen.

$$2H_{ad}^* \rightleftharpoons H_2 + 2*$$

In both the cases, the HER process starts with the adsorption of hydrogen onto the catalytic active surface (Volmer reaction) followed by the evolution of molecular hydrogen either by desorption (Heyrovsky reaction) or by the dissociative desorption (Tafel reaction) process. One can propose the reaction mechanism following by a particular catalyst from the involved rate determining step (RDS) that can be derived from the Tafel slopes. These slopes represent the intrinsic nature of a particular electrocatalysts and their magnitude provides information to recognize the reaction mechanism. The slopes are generally derived from the polarization curves by plotting the log of reduction current density against the overpotential (called the Tafel plot). It has been revealed that for a Tafel RDS the slope is ~29 mV/dec, whereas for the Heyrovsky and Volmer RDS, the slopes are ~39 and ~118 mV/dec. respectively. But generally in acid mediated hydrogen evolution reaction, the Heyrovsky step is considered as the rate determining step and Volmer step in case of the alkaline medium (Anantharaj et al. 2016; Kong et al. 2013).

Thus, the first step of the reaction involves the generation of active hydrogen species (H*) and its adsorption to the catalytic surface (Volmer step) followed by desorption as molecular hydrogen (H$_2$; by either Tafel or Heyrovsky step). The ease of adsorption as well as desorption have equal importance to monitor the reaction kinetics. Therefore the standard free energy required for adsorption (ΔG_H) of hydrogen onto the catalyst surface has been used to evaluate the interaction strength of active hydrogen with the electrode surface. Generally the material activity with respect to the value of ΔG_H is represented in form of Volcano plot (Fig. 3.2). The strong interaction of H* on the electrode surface ($\Delta G_H > 0$) slow down the Volmer step leading to a limited turnover rate. On the other hand, the weak interaction ($\Delta G_H < 0$) speed up the Volmer step and hampers harms the Tafel or Heyrovsky reaction steps.

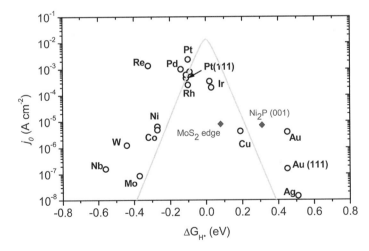

Fig. 3.2 A volcano plot of experimentally measured exchange current density as a function of the DFT-calculated Gibbs free energy of adsorbed atomic hydrogen. The simple kinetic model proposed by Nørskov and coworkers to explain the origin of the volcano plot is shown as the solid lines. (Reproduced with permission from Nørskov et al. 2005)

An optimal catalytic activity can be achieved by a catalyst having moderate (nearly zero ΔG_H) value of binding energy with the reactive intermediates (Greeley et al. 2006; Nørskov et al. 2005). Therefore the Pt based materials having nearly zero ΔG_H assumed to be the state of the art material for HER. Because of strong metal$-OH_{ad}$ interaction and requirement of extra energy for water dissociation, in alkaline medium extra overpotential is needed to furnish the HER in comparison to that of acidic medium.

3.3 Key Parameters for Catalyst Evaluation

3.3.1 Overpotential

As discussed in the above section, the electrochemical splitting of water involves two crucial steps i.e. oxygen evolution reaction and hydrogen evolution reaction at the anode and cathode of the electrolyzer respectively. Actually these reactions are not thermodynamically feasible in standard conditions of temperature and pressure and needs some extra energy. According to Nernst equation, a standard potential (E^0) of oxidation of water is 1.23 V (vs. NHE) and for water reduction 0.00 V (vs. NHE).

$$HER : H^+\left(aq\right)+2e^- \rightarrow H_2\left(g\right), E^0_{HER} = 0.00\,V$$

$$OER : 2H_2O\left(l\right) \rightarrow O_2\left(g\right)+4H^+\left(aq\right)+4e^-, E^0_{OER} = 1.23\,V$$

These are the theoretical calculations and in real electrolysis that depends on the concentration of OH^- or H^+ ions or the pH of the electrolyte. Because of the poor energy efficiency and sluggish reaction kinetics, the reaction rarely initiates at these standard potentials. Therefore some extra energy is needed to overcome the reaction to overcome reaction activation barrier and to start the HER/OER process called overpotential (η). It is the difference between the experimentally observed and thermodynamically determined potentials of a particular electrochemical reaction. The calculation of η is strictly needed to evaluate the catalytic performance of a catalyst for water electrolysis. The material having higher value of overpotential seems to have lower activity for the electrochemical water splitting. The activation of reaction, series resistance and the diffusion of charge carriers make the origin of this overpotential. Among these three, the overpotential associate with the activation of reaction is directly related to the catalytic activity of the catalyst. Therefore the evaluation of η should be carried out more precisely to access the catalytic behaviour of a material. The value of this overpotential can be minimized by choosing a suitable electrocatalyst with preferred surface morphology. Whereas the overpotential due to the charge carrier diffusion can be minimized by continuously rotation the electrode (on which the catalyst is taken) during the analysis. Meantime, there exists an ohmic resistance (mainly because of the uncompensated resistance of ionic conduction among the working and reference electrode in electrolyte) causing an ohmic drop (iR) in the observed potential leading to the resistance overpotential. Therefore, demanding further correction to the potential-current density (E–J) plot. This can be done by measuring either the solution resistance of the electrolyte using the standard electrochemical impedance spectroscopy (EIS) or by the current inter-rupt method (Cooper and Smith 2006). Generally the intrinsic catalytic property of a designed electrocatalyst can only be studied after subtracting this ohmic drop from the observed potential. In this regard the applied potential required to drive the HER/OER can be expressed as per the following equation,

$$For\ HER : E_{HER} = E^0_{HER} + iR + \eta_{HER}$$

$$For\ OER : E_{OER} = E^0_{OER} + iR + \eta_{OER}$$

Various electroactive materials (called electrocatalysts) have been developed to minimize this overpotential values. Specifically the electrocatalytic efficiency of a material can be determined by measuring the overpotential needed to achieve a particular amount of current density. The potential needed for 10 mA/cm^2 has been assumed as a standard metric in solar water splitting systems (under illumination of 1 sun) and this current density expected in a 12.3% efficient solar hydrogen device (Benck et al. 2014; Gao et al. 2019). Therefore the overpotential required for an electrocatalyst to achieve 10 mA/cm^2 (η_{10}) is assumed to be a standard process to evaluate its efficiency and to compare with other reported materials. Smaller the value of η_{10} more efficient the electrode material will be. Also in some of the more active electrocatalysts, the overpotential at higher current densities i.e. 50 mA/cm^2 (η_{50}) and 100 mA/cm^2 (η_{100}) are also used as standard metrics for activity compari-son (Chen et al. 2013; Gong et al. 2013; Lu et al. 2014). Again to achieve the true kinetic current of an electroactive material, the capacitive background current origi-nated from the adsorption/desorption of ions on the electrode material is suggested to eliminate. In case of OER this can be performed by subtracting the average capacitive current obtained from the forward and backward scans by assuming the symmetric background capacitive currents. Further, the activity of the electroactive materials can be evaluated by using different parameters i.e. Tafel slope, turnover frequency, electrochemical accessible surface area and the amount of active sites in a particular electrocatalyst etc. as discussed in the following sections.

3.3.2 Tafel Slope

Tafel slope provides the insights into the reaction mechanism of a catalysis process (HER or OER) on the electrocatalyst surface. It is generally calculated to study the intrinsic properties of a particular electrocatalyst. The Tafel plot shows the relation between the iR-corrected overpotential and current density as per the following equation,

$$\frac{d \log(j)}{d\eta} = \frac{2.303\,RT}{\alpha nF}$$

It can be clearly visible that the slope is inversely proportional to the charge transfer coefficient (α) keeping other parameters (R: gas constant, T: temperature, n: number of electrons involved in that reaction, F: Faraday constant) as constant.

$$\eta = a + b\log j$$

Here η, a, b and j are the overpotential, Tafel constant, Tafel slope and current density respectively. The Tafel plot can be derived from the polarization curves by plotting log of current density ($\log j$) against overpotential (η). Thereafter by fitting the linear portion (the linear portion indicates the onset potential of the OER/HER) of the plot, the Tafel slope can be obtained (Smith et al. 2013). Since Tafel slope is inversely proportional to charge transfer coefficient, so it has been considered as the preliminary parameter taken for the performance evaluation of a particular electrocatalyst. As discussed in the above section, the HER proceeds via two steps as Volmer step (adsorption of active hydrogen) followed by either Tafel or Heyrovsky steps (for desorption in form of molecular hydrogen). It has been stated that the theoretical Tafel slopes of 118, 39 and 29 mV/dec. are assigned to the Volmer, Heyrovsky and Tafel reaction pathways (Fletcher 2012). For example, the Tafel slope of 30 mV/dec. is observed for HER on the Pt surface in 0.5 M H_2SO_4 demonstrating the reaction proceeds via the Volmer-Tafel process with the RDS of Tafel step (Li et al. 2011).

The exchange current density (j_0) is the current density observed at an equilibrium position of a reaction (it may be HER or OER) where both the cathodic and anodic currents are equal. It can be read out from the intersection by extrapolating the linear Tafel region to the lateral axis (x-axis of the Tafel plot) (Fig. 3.3). The value of which can be used to describe the physical/surface properties, structure and other factors that influence the charge transfer at the electrode/electrolyte interfaces (Anantharaj et al. 2018). The value of j_0 depends on the catalytic active surface present on the electrocatalyst. For instance, highly active electrocatalysts give higher exchange current density compare to other having less activity.

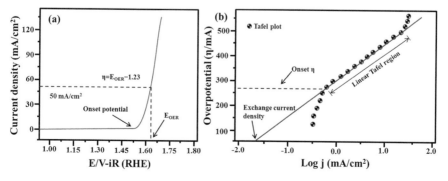

Fig. 3.3 Schematic presentation of a (**a**) linear sweep voltammogram and (**b**) Tafel plot showing the onset potential and exchange current density

3.3.3 Electrochemical Active Surface Area and Current Normalization

Generally, the observed oxidation (OER)/reduction (HER) currents are normalized either by the mass (m) of the loaded catalyst (weight specific current with unit A/gm) or by the geometrical surface area ($S_{geo.}$) of the working electrode on which the catalyst is taken (area specific current with unit A/cm^2) or the geometric surface area of the catalyst itself (S_{ECSA}; with unit A/cm^2) (Anantharaj et al. 2016). Among all of the methods, the normalization by surface area of the working electrode is widely adopted, but it unable to present the intrinsic catalytic property of the material. Moreover, variable potential for a particular current density is observed with different catalyst loading. But if the loading is 100% i.e. if it fully covers the electrode surface by forming a monolayer of catalyst, then this method of normalization can be used. Actually this is meant for a planar electrode surface (metallic foils, glassy carbon plate/disc etc. and not for metal/carbon foams, carbon cloth etc.) and is more helpful to compare with the reported literatures and to correlate with the theoretical current density values. But inconsistencies in loading amount will give variable results. In case of lower loading, more catalytically inactive electrode surface will be exposed to the electrolyte and the S_{geo} will not give the actual value. Since the reactions are surface phenomenon, so on higher loading the layer of catalyst present on the surface participated in the reaction excluding the bulk material. Therefore the catalyst loading should be optimized and consistency should be maintained throughout the work to minimize the error and also extreme care should be taken to form monolayer of catalyst.

The actual surface area of the catalyst can be derived either by *in situ* or by *ex situ* surface area analysis method. The *in situ* method can be performed either by the gas adsorption study (Brunauer-Emmet-Teller; BET) or by measuring the particle size with the help of microscope (scanning electron microscope/scanning probe microscope). Though both the method gives the average surface area of the catalyst, but all the surfaces need not be electro catalytically active. Therefore percentage of accuracy is found to be less. On the other hand, the *in situ* method is much more sensitive to the catalyst loading and can give the intrinsic property of the catalyst. This can be done by measuring the electrochemical surface area (ECSA) from the obtained double layer capacitance as discussed elaborately in the following section. The values of ECSA may be influenced by the surface roughness or porous nature of the electrocatalyst surface. Still the actual ECSA value cannot be determined because of the unclear capacitive behaviour. Instead the double layer capacitance (C_{dl}) can be obtained from the cyclic voltammograms of the respected materials. It can be derived either from the (a) static cyclic voltammograms recorded at different sweep rates in a non-Faradic potential region or from the (b) frequency dependent electrochemical impedance spectroscopy. At the time of the C_{dl} measurement, the selection of the potential range is very crucial. One should take the range of potential in which no apparent Faradic processes taking place and the capacitance value arises only because of the double layer formation at the electrode/electrolyte

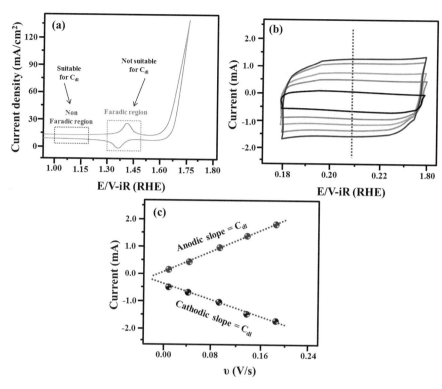

Fig. 3.4 The schematic presentation of (**a**) cyclic voltammogram showing the suitable position to consider for the C_{dl} calculation, (**b**) cyclic voltammograms in a non-Faradic region of the potential window at different scan rates and the (**c**) plot of cathodic and anodic current vs. the scan rate

interface (Fig. 3.4). The potential range is generally 0.1 V centered at the open circuit potential (OCP) of the system. Here the charging current (i_c; that obtained from the cyclic voltamograms) is the product of the scan rate (v) and double layer capacitance (C_{dl}) as per the following equation,

$$i_c = v C_{dl}$$

Therefore a plot of charging current against sweep rate gives a straight line with a slope equals to the double layer capacitance value.

In a particular work, Dutta and Samantara et al. have observed a C_{dl} value of 5.6 mF for nickel phosphide sample in 1 M KOH electrolyte (Dutta et al. 2018). In another work, Behera's groups have obtained a C_{dl} value of 0.969 mF/cm^2 for VS$_2$ nanostructures in 0.1 M H$_2$SO$_4$ solution (Das et al. 2018).

Also the value of C_{dl} can be derived from the electrochemical impedance spectrum in the similar non-Faradic potential region. Here a frequency dependent complex impedance spectrum is need to record by applying a sinusoidal potential. For example, the Jaramilo group have traced the Nyquist impedance plot (the plot of

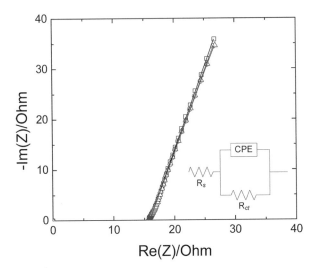

Fig. 3.5 Representative Nyquist plots for an electrodeposited NiO_x catalyst in 1 M NaOH at −0.1 V (black open square), −0.05 V (red open circle), and 0 V (blue open triangle) vs. SCE measured from EIS in the frequency range 100 kHz to 100 Hz. These potentials fall in a potential region in which no Faradaic processes are observed. The solid lines are the fits to the data using the simplified Randles circuit shown in the inset. (Reproduced with permission from McCrory et al. 2013)

imaginary vs. real impedance) in a non-faradic region within the frequency range of 100 Hz to 100 kHz (for electrodeposited NiO_x) and approximated by the modified Randle's circuit as shown in the following Fig. 3.5,

Here the R_s, R_{ct} and CPE correspond to the solution resistance, charge transfer resistance and constant phase element (CPE; double layer capacitance) of the system. The CPE is related to the frequency (ω) of the applied potential as per the following equation, (Brug et al. 1984; McCrory et al. 2013)

$$Z_{CPE} = \frac{1}{Q_0 \left(i\omega\right)^{1-\alpha}}$$

Here, I is $(-1)^{1/2}$ and Q_0 is a constant having value $1 \geq a \geq 0$. In this case, the Q_0 is related to the C_{dl} as per the following equation,

$$C_{dl} = \left[Q_0 \left(\frac{1}{R_s} + \frac{1}{R_{ct}} \right)^{(a-1)} \right]^{1/a}$$

For a = 1, the CPE acts as a pure capacitor and the value of C_{dl} is same as the value of Q_0. But for a = 0, the CPE behaves like a pure resistor and the C_{dl} cannot be defined. Thereafter the ECSA can be calculated from the C_{dl} as per the following equation,

$$ECSA = \frac{C_{dl}}{C_s}$$

Here C_s is the specific capacitance of an atomically smooth flat catalytic surface and the value assumed to be in a range of 20–60 $\mu F/cm_{ECSA}^2$ (Ma et al. 2014; Tang et al. 2015; Zhang et al. 2017). The ECSA value obtained from these two methods are not identical (standard error of ±15%) giving potential error. Thereby the normalization of current with respect to the geometrical surface area of the electrode is preferable to compare with the reported literature as well as with the theoretical current densities. Further the roughness factor (R_f; active site per unit surface area) can be obtained from the ECSA by simply dividing the geometrical surface area of the electrode as per the following equation,

$$R_f = \frac{ECSA}{S_{geo.}}$$

3.3.4 Number of Active Sites and Turnover Frequency

The number of hydrogen/oxygen molecules evolved per second per active site of an electrocatalyst is termed as the turnover frequency (TOF) (Costentin et al. 2012; Zhao et al. 2018). It is necessary to determine the activity of the reactive sites of the electrode material taken for a particular study (Ting et al. 2016). Further the number of active sites (n) present on the catalyst surface determines the value of TOF which can be obtained using either the copper underpotential deposition method or from the cyclic voltammograms. The evaluation of TOF by knowing the value of n can be as follows,

$$TOF = \frac{Number\ of\ H_2\ /\ O_2\ molecules\ evolved\ per\ second}{Number\ of\ active\ sites} \tag{3.1}$$

For example, in case of HER,

- The number of H_2 molecules evolved per second is,

$$H_2\ evolved\ per\ second = \frac{i}{96485\,C\,mol^{-1}} \times \frac{6.022 \times 10^{23}\,e^-\,mol^{-1}}{2e^-\ per\ H_2} = x \tag{3.2}$$

"i" is the current in amperes at a certain overpotential

- Number of active sites can be determined from the charge that obtained from the recorded cyclic voltammograms by assuming a one electron process for the oxidation (Merki et al. 2011).

$$Number\ of\ active\ sites = \frac{Q_{cv}}{96485\,Cmol^{-1}} \times \frac{6.022 \times 10^{23}\,e^{-1}mol^{-1}}{1e^{-}\,per\,active\,site\,oxidized} = y \qquad (3.3)$$

Then by putting these values in the above equation,

$$TOF = \frac{x}{y} = z\,s^{-1} \qquad (3.4)$$

Since four electrons are needed to evolve one oxygen molecule, so the integer 2 in the denominator of Eq. (3.2) is substituted by 4 in case of the oxygen evolution reaction. Simply the above equation can be presented as,

$$\left(TOF\right)_{HER} = \frac{i}{2Fn}$$

$$\left(TOF\right)_{OER} = \frac{i}{4Fn}$$

Here, i, F and n are the current of linear sweep voltammograms (A), Faraday constant (96485.3 C/cm^2) and the number of active sites (moles).

3.3.5 Faradic Efficiency

It is also known as the Coulombic efficiency that is an important technical index in the electrochemical reactions. Basically it measures the ratio of electron participating in a particular electrochemical reaction to the number of electron supplied. In case of OER or HER, the Faradic efficiency (FE) is defined as the number of molecular oxygen/hydrogen evolved experimentally to the theoretical values (Shi and Zhang 2016; Zou and Zhang 2015). The FE of a particular electrocatalyst can be calculated either by electrochemical method (Method-1) or by quantification of the evolved gases using different techniques (Method-2).

Method-1 The first method of FE calculation involves the use of rotating ring disk electrode (RRDE). In this case the disk (glassy carbon) is drop casted with the catalyst material which is surrounded with a platinum ring. Here the disk swept over the optimized experimental potential window keeping the ring at a constant potential. The ring potential should be noted before and is strictly depends on the pH of the electrolyte taken. The disk electrode/catalyst catalyzes the OER, whereas the ring reduces the evolved oxygen to H_2O_2. Before the measurement, the collection efficiency of RRDE should be determined through the Fe^{2+}/Fe^{3+} redox probe's response at different rotation rates. Then the FE is calculated as per the following equation,

$$FE = \frac{I_R n_D}{I_D n_R N_{CL.}}$$

Here, I_R, I_D, n_R and n_D are the ring and disc current, the number of electrons transferred at the ring and disc respectively. N_{CL} is the predetermined collection efficiency of the electrode.

Jaramillo's group has evaluated the faradic efficiency for oxygen evolution reaction using this method (McCrory et al. 2013). They have coated the disc with NiO$_x$ and performed the measurement in 1 M NaOH electrolyte at 1600 rpm (Fig. 3.6). They found the FE of ≥ 0.9 from measuring the I_D (1.95 mA) and N_{CL} (0.19) keeping I_R constant. Generally three different experiments are recommended to carry out and present the mean FE values. Although this technique is broadly adopted for the FE evaluation of many of the new HER as well as OER electrocatalysts, but still some highly sensitive techniques evolution is required for better understanding. On the other hand the theoretical value can be predicted by integrating the electrolysis (galvanostatic or potentiostatic) plot. Because of the possible side reactions, the efficiency always lies down 100%.

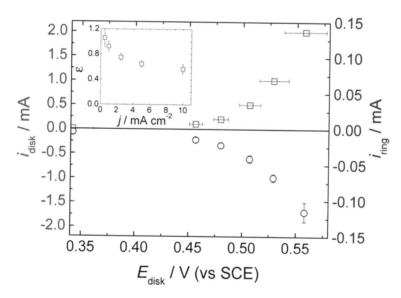

Fig. 3.6 RRDE measurements for the OER by an electrodeposited NiO$_x$ in 1 M NaOH at 1600 rpm rotation rate under 1 atm N$_2$. The disk electrode was subjected to a series of current steps, and the operating potential was measured. The dissolved oxygen generated at the disk was then reduced by 2 electrons at the surrounding Pt ring electrode. The inset shows a plot of the ratio of the ring and disk current normalized for collection efficiency and the number of electrons in the ring and disk reactions as a function of the disk current. Note that at higher disk currents, the measured current ratio deviates from the expected value of 1. This is likely due to local O$_2$- saturation and bubble formation at the disk electrode at higher current densities; only dissolved O$_2$ can be collected at the ring electrode. Faradaic efficiency measurements were calculated at id = 0.2 mA, which is j$_d$ = 1 mA/cm^2 per geometric area. (Reproduced with permission from McCrory et al. 2013)

Method-2 This is a generalized method and broadly adopted to measure the FE for both the HER and OER. Primarily the evolved gas is quantified and the ratio of experimental quantity to the theoretical value gives the FE. Here the key point is to quantify the gases during the course of or after the completion of reaction (chrono-amperometric or chronopotentiometric analysis). Generally the quantification is carried out either by following the gas displacement or chromatography or by using the spectroscopic technique (Shi and Zhang 2016). In first case, water filled inverted burette (well labeled) is connected to the electrochemical cell, during the course of reaction the evolved gases will be collected and quantified by displacement of water (Dutta et al. 2018). The second one involves the integration of a gas chromatograph to cell that quantify the gases evolved directly. On the other hand, the spectroscopic technique (third technique) can only be applied for OER that involves the excitation of evolved oxygen (from triplet to singlet state) and allow relaxing by fluorescence. The intensity of the fluorescence quantifies the oxygen gas. The use of any of these three techniques depends on the type of catalyst used, electrochemical cell dimension, types of electrode and most importantly the resources availability.

3.3.6   Mass Activity

Although the noble metal based electrocatalysts shows excellent performance towards HER and OER but their high cost and scarce reserve restricts the scalable catalyst preparation and commercialization. Therefore many efforts are made to achieve more catalytic performance with reduced use of these noble metals. In most of the reports, their performance has been compared with respect to the overpotential required for a particular area specific current, simply dividing the geometrical surface area of the electrode to the observed cathodic/anodic current. It has been observed that, all the noble metal based catalysts require nearly same overpotential to achieve the bench marked current density. However, as the catalyst made up of porous material, this type of normalization may not useful for activity comparison. Therefore the gravimetric normalization assumed to be more suitable, which can be carried out by dividing the mass of catalyst (m) taken to the observed current (I) in a particular study. This is called as the mass activity of the material and denoted by A/g.

$$Mass\ activity = \frac{I\left(A\right)}{m\left(gm\right)}$$

3.3.7   Long Cycle Life

The stability or durability is an important parameter of an electrocatalyst for its practical application and commercialization. It measures the catalytic performance of a catalyst over the given period of time. Although there are a number of techniques used to evaluate the durability of an electrocatalyst but generally the chronoamperometric and chronopotentiometric techniques are used. The former involves the measurement of variation in current density over a period of time at a particular overpotential and the latter one is the measure of change in overpotential over the time period at constant current density. It can also be evaluated by continuous cycling within the optimized potential window at a particular sweep rate. Normally, the linear sweep voltammograms are recorded before and after the durability test which may show a change in overpotential describing the performance degradation over the time period. The change in surface morphology and crystal structure/phase is observed after the durability test which may be accessed by collecting their microscopic images, diffraction pattern and elemental analysis using different advanced characterization techniques.

References

Anantharaj, S., Ede, S. R., Karthick, K., Sam Sankar, S., Sangeetha, K., Karthik, P. E., & Kundu, S. (2018). Precision and correctness in the evaluation of electrocatalytic water splitting: Revisiting activity parameters with a critical assessment. *Energy & Environmental Science, 11*, 744–771.

Anantharaj, S., Ede, S. R., Sakthikumar, K., Karthick, K., Mishra, S., & Kundu, S. (2016). Recent trends and perspectives in electrochemical water splitting with an emphasis on sulfide, selenide, and phosphide catalysts of Fe, Co, and Ni: A review. *ACS Catalysis, 6*, 8069–8097.

Benck, J. D., Hellstern, T. R., Kibsgaard, J., Chakthranont, P., & Jaramillo, T. F. (2014). Catalyzing the Hydrogen Evolution Reaction (HER) with molybdenum sulfide nanomaterials. *ACS Catalysis, 4*, 3957–3971.

Betley, T. A., Wu, Q., Van Voorhis, T., & Nocera, D. G. (2008). Electronic design criteria for O–O bond formation via metal–Oxo complexes. *Inorganic Chemistry, 47*, 1849–1861.

Bockris, J. O., & Otagawa, T. (1984). The Electrocatalysis of oxygen evolution on Perovskites. *Journal of the Electrochemical Society, 131*, 290–302.

Brug, G. J., van den Eeden, A. L. G., Sluyters-Rehbach, M., & Sluyters, J. H. (1984). The analysis of electrode impedances complicated by the presence of a constant phase element. *Journal of Electroanalytical Chemistry, 176*, 275–295.

Chen, S., Duan, J., Jaroniec, M., & Qiao, S. Z. (2013). Three-dimensional N-doped graphene hydrogel/NiCo double hydroxide electrocatalysts for highly efficient oxygen evolution. *Angewandte Chemie International Edition, 52*, 13567–13570.

Cooper, K. R., & Smith, M. (2006). Electrical test methods for on-line fuel cell ohmic resistance measurement. *Journal of Power Sources, 160*, 1088–1095.

Costentin, C., Drouet, S., Robert, M., & Savéant, J.-M. (2012). Turnover numbers, turnover frequencies, and overpotential in molecular catalysis of electrochemical reactions. Cyclic voltammetry and preparative-scale electrolysis. *Journal of the American Chemical Society, 134*, 11235–11242.

Das, J. K., Samantara, A. K., Nayak, A. K., Pradhan, D., & Behera, J. N. (2018). VS2: An efficient catalyst for an electrochemical hydrogen evolution reaction in an acidic medium. *Dalton Transactions, 47*, 13792–13799.

De Faria, L. A., Boodts, J. F. C., & Trasatti, S. (1996). Electrocatalytic properties of ternary oxide mixtures of composition Ru0.3Ti(0.7−x)CexO2: Oxygen evolution from acidic solution. *Journal of Applied Electrochemistry, 26*, 1195–1199.

Dutta, A., Mutyala, S., Samantara, A. K., Bera, S., Jena, B. K., & Pradhan, N. (2018). Synergistic effect of inactive iron oxide core on active nickel phosphide shell for significant enhancement in oxygen evolution reaction activity. *ACS Energy Letters, 3*, 141–148.

Edmonds, T., & McCarroll, J. J. (1978). Impact of surface physics on catalysis. In Gates, B, Knoezinger H (eds.), *Topics in surface chemistry* (1st ed.). Springer US, Boston, MA, pp 261–290.

Fletcher, S. (2012). Physical electrochemistry. Fundamentals, techniques, and applications by Eliezer Gileadi. *Journal of Solid State Electrochemistry, 16*, 1301–1301.

Gao, Q., Zhang, W., Shi, Z., Yang, L., & Tang, Y. (2019). Structural design and electronic modulation of transition-metal-carbide electrocatalysts toward efficient hydrogen evolution. *Advanced Materials, 31*, 1802880.

Gerken, J. B., McAlpin, J. G., Chen, J. Y. C., Rigsby, M. L., Casey, W. H., Britt, R. D., & Stahl, S. S. (2011). Electrochemical water oxidation with cobalt-based electrocatalysts from pH 0–14: The thermodynamic basis for catalyst structure, stability, and activity. *Journal of the American Chemical Society, 133*, 14431–14442.

Gong, M., Li, Y., Wang, H., Liang, Y., Wu, J. Z., Zhou, J., Wang, J., Regier, T., Wei, F., & Dai, H. (2013). An advanced Ni–Fe layered double hydroxide electrocatalyst for water oxidation. *Journal of the American Chemical Society, 135*, 8452–8455.

Goodenough, J. B., Manoharan, R., & Paranthaman, M. (1990). Surface protonation and electrochemical activity of oxides in aqueous solution. *Journal of the American Chemical Society, 112*, 2076–2082.

Greeley, J., Jaramillo, T. F., Bonde, J., Chorkendorff, I., & Nørskov, J. K. (2006). Computational high-throughput screening of electrocatalytic materials for hydrogen evolution. *Nature Materials, 5*, 909–913.

Halck, N. B., Petrykin, V., Krtil, P., & Rossmeisl, J. (2014). Beyond the volcano limitations in electrocatalysis – Oxygen evolution reaction. *Physical Chemistry Chemical Physics, 16*, 13682–13688.

Hammer, B. (2006). Special sites at Noble and late transition metal catalysts. *Topics in Catalysis, 37*, 3–16.

Haumann, M., Liebisch, P., Müller, C., Barra, M., Grabolle, M., & Dau, H. (2005a). Photosynthetic O_2 formation tracked by time-resolved X-ray experiments. *Science (80-.), 310*, 1019 LP–1021.

Haumann, M., Müller, C., Liebisch, P., Iuzzolino, L., Dittmer, J., Grabolle, M., Neisius, T., Meyer-Klaucke, W., & Dau, H. (2005b). Structural and oxidation state changes of the photosystem II Manganese complex in four transitions of the water oxidation cycle (S0 → S1, S1 → S2, S2 → S3, and S3,4 → S0) characterized by X-ray absorption spectroscopy at 20 K and room temperature. *Biochemistry, 44*, 1894–1908.

Hibbert, D. B. (1980). The electrochemical evolution of O2 on NiCo2O4 in 18O-enriched KOH. *Journal of the Chemical Society, Chemical Communications*, 202–203. https://doi.org/10.1039/C39800000202

Hong, W. T., Risch, M., Stoerzinger, K. A., Grimaud, A., Suntivich, J., & Shao-Horn, Y. (2015). Toward the rational design of non-precious transition metal oxides for oxygen electrocatalysis. *Energy & Environmental Science, 8*, 1404–1427.

Kong, D., Cha, J. J., Wang, H., Lee, H. R., & Cui, Y. (2013). First-row transition metal dichalcogenide catalysts for hydrogen evolution reaction. *Energy & Environmental Science, 6*, 3553–3558.

Koper, M. T. M., & van Santen, R. A. (1999). Interaction of H, O and OH with metal surfaces. *Journal of Electroanalytical Chemistry, 472*, 126–136.

Kosmulski, M. (2009). pH-dependent surface charging and points of zero charge. IV. Update and new approach. *Journal of Colloid and Interface Science, 337*, 439–448.

Li, Y., Wang, H., Xie, L., Liang, Y., Hong, G., & Dai, H. (2011). MoS2 nanoparticles grown on graphene: An advanced catalyst for the hydrogen evolution reaction. *Journal of the American Chemical Society, 133*, 7296–7299.

Lu, Z., Xu, W., Zhu, W., Yang, Q., Lei, X., Liu, J., Li, Y., Sun, X., & Duan, X. (2014). Three-dimensional NiFe layered double hydroxide film for high-efficiency oxygen evolution reaction. *Chemical Communications, 50*, 6479–6482.

Ma, T. Y., Dai, S., Jaroniec, M., & Qiao, S. Z. (2014). Metal–organic framework derived hybrid Co3O4-carbon porous nanowire arrays as reversible oxygen evolution electrodes. *Journal of the American Chemical Society, 136*, 13925–13931.

Man, I. C., Su, H.-Y., Calle-Vallejo, F., Hansen, H. A., Martínez, J. I., Inoglu, N. G., Kitchin, J., Jaramillo, T. F., Nørskov, J. K., & Rossmeisl, J. (2011). Universality in oxygen evolution electrocatalysis on oxide surfaces. *ChemCatChem, 3*, 1159–1165.

Mavros, M. G., Tsuchimochi, T., Kowalczyk, T., McIsaac, A., Wang, L.-P., & Van Voorhis, T. (2014). What can density functional theory tell us about artificial catalytic water splitting? *Inorganic Chemistry, 53*, 6386–6397.

McCrory, C. C. L., Jung, S., Peters, J. C., & Jaramillo, T. F. (2013). Benchmarking heterogeneous Electrocatalysts for the oxygen evolution reaction. *Journal of the American Chemical Society, 135*, 16977–16987.

Merki, D., Fierro, S., Vrubel, H., & Hu, X. (2011). Amorphous molybdenum sulfide films as catalysts for electrochemical hydrogen production in water. *Chemical Science, 2*, 1262–1267.

Mun, B. S., Watanabe, M., Rossi, M., Stamenkovic, V., Markovic, N. M., & Ross, P. N. (2005). A study of electronic structures of Pt3M (M=Ti,V,Cr,Fe,Co,Ni) polycrystalline alloys with valence-band photoemission spectroscopy. *The Journal of Chemical Physics, 123*, 204717.

Noh, J. S., & Schwarz, J. A. (1989). Estimation of the point of zero charge of simple oxides by mass titration. *Journal of Colloid and Interface Science, 130*, 157–164.

Nørskov, J. K., Bligaard, T., Logadottir, A., Bahn, S., Hansen, L. B., Bollinger, M., Bengaard, H., Hammer, B., Sljivancanin, Z., Mavrikakis, M., Xu, Y., Dahl, S., & Jacobsen, C. J. H. (2002). Universality in heterogeneous catalysis. *Journal of Catalysis, 209*, 275–278.

Nørskov, J. K., Bligaard, T., Logadottir, A., Kitchin, J. R., Chen, J. G., Pandelov, S., & Stimming, U. (2005). Trends in the exchange current for Hydrogen evolution. *Journal of the Electrochemical Society, 152*, J23–J26.

Patrick, J. W. (2004). Handbook of fuel cells. Fundamentals technology and applications. *Fuel, 83*, 623. https://doi.org/10.1016/j.fuel.2003.09.012.

Pechenyuk, S. (1999). The use of the pH at the point od zero charge for characterizing the properties of oxide hydroxides. *Russian Chemical Bulletin, 48*, 1017–1023.

Rossmeisl, J., Logadottir, A., & Nørskov, J. K. (2005). Electrolysis of water on (oxidized) metal surfaces. *Chemical Physics, 319*, 178–184.

Shi, Y., & Zhang, B. (2016). Recent advances in transition metal phosphide nanomaterials: Synthesis and applications in hydrogen evolution reaction. *Chemical Society Reviews, 45*, 1529–1541.

Smith, R. D. L., Prévot, M. S., Fagan, R. D., Trudel, S., & Berlinguette, C. P. (2013). Water oxidation catalysis: Electrocatalytic response to metal stoichiometry in amorphous metal oxide films containing Iron, cobalt, and nickel. *Journal of the American Chemical Society, 135*, 11580–11586.

Suntivich, J., May, K. J., Gasteiger, H. A., Goodenough, J. B., & Shao-Horn, Y. (2011). A Perovskite oxide optimized for oxygen evolution catalysis from molecular orbital principles. *Science (80-.), 334*, 1383 LP–1385.

Tang, C., Wang, W., Sun, A., Qi, C., Zhang, D., Wu, Z., & Wang, D. (2015). Sulfur-decorated molybdenum carbide catalysts for enhanced Hydrogen evolution. *ACS Catalysis, 5*, 6956–6963.

Ting, L. R. L., Deng, Y., Ma, L., Zhang, Y.-J., Peterson, A. A., & Yeo, B. S. (2016). Catalytic activities of sulfur atoms in amorphous molybdenum sulfide for the electrochemical hydrogen evolution reaction. *ACS Catalysis, 6*, 861–867.

Trasatti, S. (1991). Physical electrochemistry of ceramic oxides. *Electrochimica Acta, 36*, 225–241.

Wohlfahrt-Mehrens, M., & Heitbaum, J. (1987). Oxygen evolution on Ru and RuO2 electrodes studied using isotope labelling and on-line mass spectrometry. *Journal of Electroanalytical Chemistry and Interfacial Electrochemistry, 237*, 251–260.

Zhang, C., Huang, Y., Yu, Y., Zhang, J., Zhuo, S., & Zhang, B. (2017). Sub-1.1 nm ultrathin porous CoP nanosheets with dominant reactive {200} facets: A high mass activity and efficient electrocatalyst for the hydrogen evolution reaction. *Chemical Science, 8*, 2769–2775.

Zhao, G., Rui, K., Dou, S. X., & Sun, W. (2018). Heterostructures for electrochemical Hydrogen evolution reaction: A review. *Advanced Functional Materials, 28*, 1803291.

Zou, X., & Zhang, Y. (2015). Noble metal-free hydrogen evolution catalysts for water splitting. *Chemical Society Reviews, 44*, 5148–5180.

Chapter 4
Electroactive Materials

Abstract Both the HER and OER are the core reactions of advanced energy conversion technologies (Fuel cell, metal-air batteries, conversion of CO2 to fuel etc.) which are the key components of the sustainable energy utilization infrastructures. But experimentally these reactions face higher activation energy barrier and require additional potential called overpotential for their completion. In order to reduce the overpotential, numerous electrocatalyst and advanced techniques for electrode fabrication were developed. This chapter covers an up to date literature survey on different types of electroactive materials (precious metals, metal oxides, chalcogenides, phosphides, supported materials etc.) and electrodes employed for electrolysis purpose.

Keywords Electrode materials · Noble metals · Nanostructures · Metal oxides · Chalcogenides · Sulphides · Selenides · Graphene CNT · Composites · Supported electrodes · Free standing electrodes

Both the HER and OER are the core reactions of advanced energy conversion technologies (Fuel cell, metal-air batteries, conversion of CO_2 to fuel etc.) which are the key components of the sustainable energy utilization infrastructures. As discussed above, the HER is a simple reaction that involves two electron process to generate H_2 from H_2O at a thermodynamic potential of 0 V (vs. RHE). On the other hand, the overall OER involves four electron transfer path to execute the effective conversion of H_2O to O_2 at a thermodynamic equilibrium potential of 1.23 V (vs. RHE). The HER and OER generally follows a Butler-Volmer reaction model and require higher overpotential. Further the reaction mechanisms are very sensitive to the crystal structure and morphology of the electrode surfaces (Kreysa et al. 2014; Mom et al. 2014). Therefore precious metals like iridium (Ir), platinum (Pt), rhodium (Rh), ruthenium (Ru) etc. are employed to achieve effective conversion to these gaseous molecules with reduced overpotential. But the high cost, limited reserve in the earth crust, operation durability and long term storage restricts the large scale applications demanding alternative materials.

© The Author(s), under exclusive license to Springer Nature Switzerland AG 2019
A. K. Samantara, S. Ratha, *Metal Oxides/Chalcogenides and Composites*,
SpringerBriefs in Materials, https://doi.org/10.1007/978-3-030-24861-1_4

Transition metals

Legend: Metal Oxides, Metal Borides, Metal Carbides, Metal Nitrides, Metal Phosphides

											5 B	6 C	7 N	8 O
											13 Al	14 Si	15 P	16 S
21 Sc	22 Ti	23 Cr	24 Cr	25 Mn	26 Fe	27 Co	28 Ni	29 Cu	30 Zn	31 Ga	32 Ge	33 As	34 Se
39 Y	40 Zr	41 Nb	42 Mo	43 Tc	44 Ru	45 Rh	46 Pd	47 Ag	48 Cd	49 In	50 Sn	51 Sb	52 Te
57 La	72 Hf	73 Ta	74 W	75 Re	76 Os	77 Ir	78 Pt	79 Au	80 Hg	81 Tl	82 Pb	83 Bi	84 Po
89 Ac	104 Rf	105 Db	106 Sg	107 Bh	108 Hs	109 Mt	110 Ds	111 Rg	112 Cp				

Fig. 4.1 Schematic presentation of the elements used to prepare non-noble metal based HER and OER electrocatalysts in the periodic table

In this regard, numerous efforts have been paid to design non-noble metal based electrocatalysts and catalysts with low precious metal content and demonstrated their effective catalysis performances. These are the metal oxides, chalcogenides, mixed oxides, noble metal composites, and hybrids with carbonaceous materials and etc. that are schematically presented in Fig. 4.1 and discussed briefly in the following sections.

4.1 Precious Metal Based Catalysts

Primarily, the precious metals were considered as the only efficient electrocatalysts for the OER. Experimentally, it has been observed that the overpotential required by the metal electrocatalysts executing the OER in acidic medium follows; Ru < Ir < Pd < Rh < Pt. Hence the Ru metal is consider as relatively best and Pt as worst for OER. But during the electrochemical measurement, at higher applied potential, surface of these metals get corrode by forming a layer of the respective oxides showing poor catalytic stability. Therefore the stability trend became just reversed to that of catalytic activity and Pt based electrocatalyst shows better stability compared to that of Ru. In fact some of the precious metals form a stable oxide layer and shows better activity towards OER. Since it has been established that the noble metals are good choice for the OER, but high cost and limited availability restricted their commercialization. Therefore research is going on to reduce the cost of these precious catalysts by reducing the content and targeting to maximize the utilization of all the atomic active centers. This may be achieved by reducing the particle size and making a uniform distribution on a suitable conductive support of higher surface area. The size reduced particles not only avails more accessible surface area but also reduce the energy barrier for the homogeneous nucleation forming more number of active sites for the oxygen evolution reaction. However the maximum performance has not been realized because of the limitations like to reduce the particle size and to make a uniform distribution over the support material. Further the combination of two metals in a core shell fashion provides remarkable catalytic activity. This additional performance can be attributed to the formation of an uncommon phase at the metal/metal interface or other active sites (Xia et al. 2014). Also the synergistic effect in the catalytic performance has been observed by alloying two or more metals. A possible alteration in the uniform metallic lattice structure takes place by the arrangement of different sized metal atoms demonstrating enhanced electrocatalytic performances (Neyerlin et al. 2009; Pei et al. 2016; Yuan et al. 2016).

Interestingly among oxide of these noble metals, the rutile stage of RuO_2 shows higher activity for OER both in the alkaline as well as acidic electrolyte. Based on the theoretical calculations, its catalytic activity is well supported by its surface structure, crystallinity, porosity and roughness factor etc. But it is found to be unstable in acidic medium and converts to dissolvable RuO_4 at higher applied anodic potential (>1.4 V) giving poorer electrocatalytic activity. Thereafter, the IrO_2 considered being the best substitute to RuO_2 showing better catalytic durability up to 2 V of applied potential. Based upon the theoretical calculations, superior electrocatalytic behavior of IrO_2 is supported by the stability of the intermediates involved in the OER mechanism. Further the stability of intermediates was improved by forming mixed metal oxides. In a particular report, Tang et al. have observed better stability of the intermediates in $SrIrO_3$ (100) compared to the IrO_2 (R. Tang et al. 2016). Although in both the cases, the active catalytic site is Ir^{4+}, but the enhanced oxygen coordination environment provides more stability in the composite. Also the time resolved UV-Vis. spectroscopy, in situ X-ray absorption spectroscopy (XAS) and X-ray photoelectron spectroscopy (XPS) studies clearly identify the intermediate conducting OER Ir^{4+} or Ir^{5+} (Hüppauff 1993; Nahor et al. 1991; Sanchez Casalongue et al. 2014; Sardar et al. 2014). Further various synthetic methodologies have been designed for the preparation of Ir-Ru mixed metal oxides and introduction of other non-precious metals to reduce the use of these precious metals contents (Kötz and Stucki 1985; Mamaca et al. 2012). Specifically, Yeo et al., have prepared a ternary oxide by introducing tantalum into the binary oxide of Ir-Ru (Yeo et al. 1981). In another work, Hutchings et al. have prepared tin containing Ir-Ru ternary oxides that show better cyclic durability preserving the catalytic activity (Hutchings et al. 1984).

In case of HER, precious metal Pt is assumed to be most efficient electrocatalyst, but the limited reserve and high cost restrict their commercialization. Therefore strategies have been designed to reduce the usage of Pt and to achieve better catalytic performance as well as operational durability. The first attempt in this regard was taken by alloying the Pt with a suitable low precious 3d transition metal (like Pd and Ni) or by depositing a thin layer of Pt onto a suitable substrate. Here the substrate not only acts as a support to make the atomic layer of Pt but also can able to optimize the d-band center and ΔG_H of Pt promoting the better performance of the corresponding bimetallic system. The configuration with the 3d metal on the top-most layer of Pt(111) surface and in the subsurface region make the d-band center to move close to the Fermi level and away from the Fermi level respectively. Such phenomenon observed to be realized in case of the depositing monolayers of Pt or Pd (by underpotential deposition methods) on to the single crystalline metal surfaces (Esposito and Chen 2011; Greeley et al. 2006; Kelly and Chen 2012; Kibler et al. 2005; Kitchin et al. 2004). These bimetallic electrocatalysts shows better catalytic performances than that of single metals. But further focus has been paid to reduce the use of these noble metals and design some alternative material having better electrocatalytic performances. On the other hand the selection of a suitable substrate is another challenge to work out. Among different transition metal based materials, the metal carbides (generally tungsten and molybdenum carbides) found to be suitable one for this purpose. From the DFT calculations, it has been revealed that a variety of metals (like Cu, Ag, Au, Pd, Pt etc.) can be efficiently deposited as monolayers onto the (0001) surface of WC (Vasić Anićijević et al. 2013). By manipulating the surface electronic structure, these surface coated tungsten carbides (even in very low content) show similar HER performance to the pure metal.

4.2   Metal Oxides/Hydroxides

In addition to these precious metal oxides, the non-precious transition metal based oxides also been thoroughly investigated towards OER. Among them, the Ni and Co based oxides possess profound electrocatalytic activity and durability in alkaline electrolyte. Particularly the CoO_x, Co_xO_y, NiO were extensively explored as efficient anode electrocatalyst for water electrolysis. The Co-based oxides are broadly studied for OER due to their better electrocatalytic activity and corrosion resistance towards the reactions in the alkaline media. The Co_xO_y i.e. Co_3O_4 shows a noticeable catalytic performance for OER and the effective active site was found to be two Co^{4+} ions connected by bridging oxos. The Oxos interacts with the water molecule during the electrolysis process. The oxo coordination significantly reduces the activation barrier and the overpotential required to execute OER (Plaisance et al. 2016). Therefore by modifying the surface electronic structure, the overpotential can be reduced thereby increasing the efficiency of an electrocatalyst. On the other hand, the amorphization of the active surface were also be used to increase the participation of catalyst to form the reaction intermediates for OER (Bergmann et al. 2015). Leng et al. have observed the formation of hydrated and carbonated cobalt hydroxides on the amorphous surface of Co_3O_4 that are more active for OER electrolysis (Leng et al. 2015). Here the Co^{4+} active site assumed to be agglomerated in a particular fashion on the catalytic surface providing enhanced performance. Also the number of active sites in Co_3O_4 depends on the ratio of Co^{3+} to Co^{4+} present on its surface. In case of crystalline materials, the number of active sites strongly depends on the crystal planes (Zhang et al. 2015). In particular, though the Co_3O_4 contains both the Co^{2+} and Co^{3+}, but the Co^{3+} crystal facet shows faster OER kinetics. Since the OER is a surface phenomenon and the surface electronic state plays a significant role in the catalytic activity, therefore various methods like doping, creation of oxygen vacancies and facet controlled synthesis of nanostructures were employed to tune the catalytic activity of the Co_3O_4 towards OER (Chen et al. 2015; N.-I. Kim et al. 2016; Y. Wang et al. 2014; Wu and Scott 2013; Zou et al. 2013). Not only the electronic structure but also the surface morphology plays a significant role in determining the catalytic efficiency of a particular electrocatalyst. The number of active site increases by tuning the electronic structure, whereas the morphology guides the channels/path for transportation of ions/electrons and reaction products. But the active sites are not directly proportionate to the surface area of the catalyst, since an optimum dimension of the catalyst can only catalyze the OER efficiently. Another factor i.e. chemical irregularity along with dimension of the nanostructured particles observed to be having importance in manipulation of OER performances. Therefore the Co_3O_4 quantum dot shows superior OER performance that may be because of their small size as well as surface irregularities. Further the doping of different atoms shows non-uniformity in the charge distribution demonstrating an improved OER performance. In a particular work, Kim et al. have studied a variety of doped Co_3O_4 and observed that the crystallite size has a direct relationship with the OER activity (N.-I. Kim et al. 2016). Smaller the crystallites, lower will be charge transfer resistance thereby showing enhanced OER performance.

Different oxides of manganese (α-MnO_2, β-MnO_2, γ-MnO_2, Mn_3O_4 and etc.) were developed and their catalytic activity has been studied. The presence of Mn^{3+} ions in a distorted crystal lattice, existence of structural disorder and variable Mn-O bond lengths originates the catalytic efficiency of these manganese oxides. The spacious 2×2 channels in alpha form of MnO_2 avail suitable space for the movement/diffusion of electrolyte ions as well as reaction products demonstrating better OER performance (Boppana and Jiao 2011). But the catalytic performance is still lag behind its counterparts (very low than expected), which may be due to their poor conductivity. By adding metallic catalysts, the electrical conductivity increases showing improved catalytic activity. Kuo et al. observed nearly six times enhanced OER activity after adding a small amount of gold nanoparticles (Au NPs) to α-MnO_2 (Kuo et al. 2015). Here the electronic interaction of Au NPs not only provides extra conductivity to the catalyst but also promote the in situ generation of active sites (Mn^{3+}) benefiting OER performance. On the other hand, narrow internal lattice structure and absence of di-μ2-oxo bridged Mn centers make the β form of MnO_2 very poor OER activity. But by optimizing the reaction conditions, Kim et al. have prepared porous β-MnO_2 nanoplates showing more accessible surface area demonstrating improved OER performance (J. Kim et al. 2016). Further the presence of both the Mn^{2+} and Mn^{3+}, make the Mn_3O_4 as a potential candidate towards OER (Guo et al. 2014; Huynh et al. 2015). Another transition metal catalyst i.e. copper oxide is cheaper compared to Co and Ni based catalysts and shows considerable catalytic activity towards OER. But this activity found to be very less than the above discussed Ni and Co- based oxides. Owing to its flexible coordination number, it is possible to synthesize various complexes of copper to tune the morphology of resulting copper oxide nanostructures for improved OER (X. Liu et al. 2016).

Although various metal oxides are developed and explored their electrocatalytic performances towards OER but the activity still remains far from the noble metal based catalysts. Further to achieve the noble metal like catalytic performances, mixed metal oxides and perovskites were developed that obviously show better performance than the individual oxides. Generally by doping a second metal to the lattice of a transition metal oxide generate a mixed metal oxide structure with an improved catalytic activity (Bau et al. 2014; Qiu et al. 2014; Smith et al. 2013). Particularly, the doping of Ni, Mn, Li and Cu in CoO_x enhance the catalytic performances. In these cases, the enhanced OER activity is mainly ascribed to the change in lower/ higher oxide transition energies and oxide work function after the substitution of metal atoms (Trasatti 1984). On the basis of theoretical observations, various types of metal oxides like perovskites, rutile and anatase oxides shows OER behaviour and their catalytic activity trend depends on the free energies of the intermediates (OOH$*$ and OH$*$) as discussed in the above section (Man et al. 2011). On changing the reaction conditions and precursors, different doped nanostructures with various surface morphologies has been developed that avails more active surfaces showing better catalytic performances. Bau et al. have synthesized nickel-iron oxide single crystal nanoparticles by refluxing the metal oleate where both the ions are present in +2 oxidation states (Bau et al. 2014). However the iron active sites observed to be responsible for oxygen evolution (the Fe^{2+} oxidized to Fe^{3+}). Since the surface structure plays an important role in catalysis, the surfaces of this metal oxide are treated with plasma or UV irradiation to develop more active sites for OER.

Further the spinels like $NiCo_2O_4$, $CuCo_2O_4$, $ZnCo_2O_4$, $CoFe_2O_4$, $NiMoO_4$, $MnCo_2O_4$ etc. shows excellent electrocatalytic behaviour for OER. Interestingly the catalytic activities of these mixed oxides are observed to be more compared to that of individual counterparts. For example, the OER activity of $NiCo_2O_4$ is more than those of NiO and Co_3O_4. This activity may be due to the non-uniform distribution of cationic charges that creates active sites for the electrocatalysis. But their poor electrical conductivity limits their catalytic performances restricting its practical application. Therefore composites of these spinels with a suitable conducting material are developed that shows an improved OER performance.

By a simple hydrothermal reaction followed by a post annealing method, Samantara et al. have prepared spinel $NiCo_2O_4$ porous nano rods (NCO NRs) and its composite with reduced graphene oxide (RG/NCO NCs) (Fig. 4.2) (Samantara et al. 2018). The one dimensional structure of the nano rods shorten the path of ion transport thus decreasing the charge transfer resistance (can be seen in the above Nyquist impedance plot of Fig. 4.2g) of the RG/NCO NCs leading to faster mass transport process. Further the porous nature of the nano rods and presence of conductive reduced graphene oxide sheets provides additional surface area availing more accessible active sites. Therefore the composite shows improved water oxidation efficiency in terms of η_{10} (overpotential required to achieve 10 mA/cm^2 current density; 313 mV), lower Tafel slope (35 mV/dec) and long term durability compared to the bare nano rods, IrO_2/C and many of the reported oxide catalysts. In another report, the same group has synthesized the metastable β-$NiMoO_4$ and observed an improved catalytic OER activity (η_{10} = 300 mV, Tafel slope = 53 mV/dec) in alkaline medium (Ratha et al. 2017). In case of the Mn-Co spinels, the ratios of Mn to Co monitor the catalytic performance by suppressing the Jahn-Teller distortion caused by the Mn^{3+}

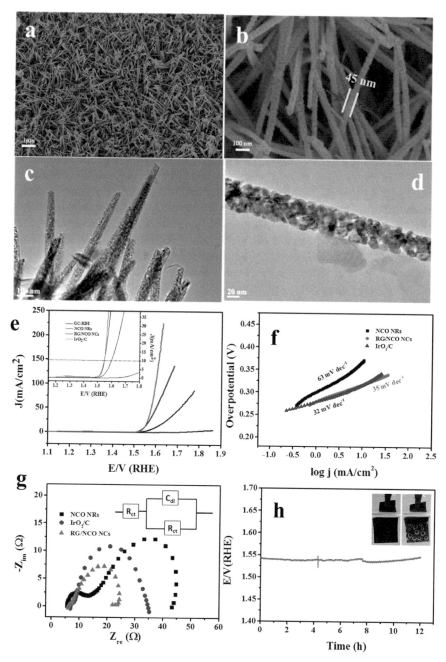

Fig. 4.2 FESEM (**a**, **b**) and TEM (**c**, **d**) images of NCO NRs at different magnification, Linear sweep voltamograms for OER by the NCO NRs and RG/NCO NCs in 1 M KOH electrolyte (**e**), corresponding Tafel plot (**f**), Nyquist impedance plot (**g**) and stability test at 10 mA/cm² current density (**h**). The inset in (**h**) is the digital photograph taken during the stability test. (Reproduced with permission from Samantara et al. 2018)

ions (Hirai et al. 2016). In another work, Li et al. have reported the catalytic activity of different transition metal iron oxide spinels of type MFe_2O_4 and concluded that the Co-based spinel ($CoFe_2O_4$) shows better OER performance (M. Li et al. 2015).

On the other hand the metal hydroxides specifically the layered metal hydroxides attracted huge attention as a promising OER electrocatalyst. They have cationic brucite like layered structure well separated by charge balancing inter layer anions. Also their open structure provides more catalytic active sites allowing the rapid diffusion of the reactants and products demonstrating excellent proton coupled electron transfer reactions. Basing upon the strength of hydrogen adsorption onto the metal surfaces ($OH_{ad}-M^{2+\delta}$ energy; Ni < Co < Fe < Mn), the 3d transition metal hydroxides show OER activity as per the trend of Mn < Fe < Co < Ni (Subbaraman et al. 2012). Specifically, the Ni based hydroxides have moderate $OH_{ad}-M^{2+\delta}$ energy thereby showing maximum catalytic activity. Therefore more focus have been paid to develop the $Ni(OH)_2$ crystal structures with various surface morphology. Although all crystal forms of $Ni(OH)_2$ (i.e. $\alpha-Ni(OH)_2$ and $\beta-Ni(OH)_2$) are catalytically active but among them the α-form shows a dominant behaviour in alkaline electrolytic condition (Gao et al. 2014). As like in case of the above discussed TMOs, the surface morphology and porous structure of metal hydroxides also plays a vital role in determining the OER performance. Li et al. have reported one dimensional Co/Ni based bimetallic hydroxides and noticed a remarkable OER performance in terms of lower overpotential, small Tafel slope with high oxidation current density (S. Li et al. 2016). Here the 1D structure shows a shorten charge conduction path availing more catalytically active sites along with the octahedral coordination states beneficiating the OER performance. Also the incorporation/doping of hydroxides with metal impurities create disorder in the metal hydroxide/metal hydr(oxy) oxide crystal structures showing an improved catalytic performance. For example, Boettcher's group have introduced Fe impurities into the crystals of γ-NiOOH and observed superior OER activity than that of the β-form (Trotochaud et al. 2012). Moreover the layer structure of the layered double hydroxides (LDHs of CoCo, NiCo, NiFe etc.) facilitates the intercalation/deintercalation of ions/water molecules providing better bulk electrolysis (Song and Hu 2014).

Whereas the tunable surface morphology and ease of preparation, the Perovskite oxides (ABO_x; A = alkaline earth/ rare earth metal and B = transition metals) and hydroxides attracts the energy researchers to explore their OER activity. Hu's group have synthesized the perovskite hydroxide of CoSn (i.e. $CoSn (OH)_6$ nanocubes) and observed excellent OER performance with $\eta_{10} = 274$ mV (Song et al. 2016). The superior activity has been ascribed to the in situ formed CoOOH hierarchical nanoporous particles (by dissolution of selective tin hydroxides). Not only the perovskites but also the double perovskites ($A_2BB'O_6$; A = large cation, B & B′ = smaller cations) found to be much more active towards OER. Diaz-Morales et al. have studied the OER activity of Ir-based double perovskites (DPs) that shows nearly threefold higher activity (in terms of Tafel slope) compared to that of IrO_2 nanoparticles in acidic medium. In this case the three dimensional network of corner shared octahedral plays a dominant role in improvement of the catalytic activity and durability (Diaz-Morales et al. 2016). Afterwards by changing the reaction conditions and precursor ratios, different research groups have designed various perovskite based oxide and hydroxides and explored their OER performance in both the alkaline as well as acidic medium (Guo et al. 2015). A significant improvement in catalytic activity can also be realized by creating crystal vacancies in perovskites materials. In this aspect, Suntivich et al. have prepared $Ba_{0.5}Sr_{0.5}Co_{0.8}Fe_{0.2}O_{3-\delta}$ (BSCF) and observed superior OER performance compared to the state of the art catalyst IrO_2 in alkaline medium (Suntivich et al. 2011). After a thorough investigation of ten transition metal oxides, the authors have derived a principle that correlate the catalytic activity with the occupancy of the 3d electron and e_g symmetry presented in a following volcano plot.

The transition metal with e_g occupancy close to unity having higher covalency of metal-oxygen bond shows better OER activity. In another report, Zhu et al. have presented the OER performance of P-doped tetragonal superstructure perovskite oxide $(SrCo_{0.95}P_{0.05}O_{3\delta})$ (Zhu et al. 2016). The accelerated OER activity is attributed to the combined effect of good electrical conductivity, existence of more O_2/O-species and P-doping in the perovskite. However the lower intrinsic electrical conductivity and incompatible hydrogen adsorption free energy limits the transition metal oxides (TMO) to electro catalyze the hydrogen evolution reaction. Since they have excellent OER performances, it is necessary to design the TMO based electrocatalysts for HER to make a bifunctional catalyst for overall water splitting. Therefore immense research effort like nanostructurization, formation of hetero structures, introduction of oxygen vacancies, composite preparation with carbon and functionalized carbon materials, doping of hetero atoms in the oxide catalysts etc. have made to enhance the hydrogen evolving efficiency of these TMO based electrocatalysts (Gong et al. 2014; Jing et al. 2018; Y. H. Li et al. 2015; H. Wang et al. 2015; Yan et al. 2015; Zhong et al. 2016). Further these TMOs are observed to be unstable in acidic media and can only catalyze the HER in alkaline electrolytic condition. In a particular work, Zhang et al. have synthesized NiO nano rods (Fig. 4.3a) with oxygen vacancies and explored the relationship among the concentration of oxygen vacancies with the HER performances (T. Zhang et al. 2018). The introduction of oxygen vacancies creates new electronic states close to the Fermi level of NiO nano rods (Fig. 4.3b).

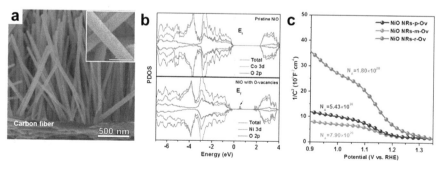

Fig. 4.3 (**a**) SEM images of NiO nanorod arrays on carbon fibers. (**b**) The projected density of states on pristine NiO and NiO with oxygen-vacancies. (**c**) Mott-Schottky plots of NiO NRs with varied amounts of oxygen-vacancies. (Reproduced with permission from T. Zhang et al. 2018)

Also a larger carrier concentration has been obtained with an optimum concentration of vacancies (as shown in the Mott-Schottky plot, Fig. 4.3c). It results a higher electron transfer rate in the oxygen vacancy enriched NiO NRs thereby enhancing the HER performances. Among other transition metal oxides, the tungsten oxide (WO_x) shows much more thermodynamically stability in acidic solution attracting the research interest in electrocatalysis. Inspired by this attractive behaviour, Li et al. have prepared oxygen vacancy enriched $WO_{2.9}$ by thermally treating the commercially available WO nanoparticles and studied the HER activity in acidic medium (Y. H. Li et al. 2015). An enhanced electrocatalytic HER performance has been achieved in these oxygen vacancy contained $WO_{2.9}$ in terms of lower Tafel slope (50 mV/dec) and overpotential required to deliver the state of the art current density ($\eta_{10} = 70$ mV) compared to that of pristine WO_3 (Tafel slope of 120 mV/dec and $\eta_{10} = 637$ mV). This experimental observation has been supported by the Density Functional Theory (DFT) calculations showing the presence of (010) and (001) as the most stable crystal facets in $WO_{2.9}$ and WO_3 respectively. Also the free energy for hydrogen adsorption on $WO_{2.9}$ (010) is nearly close to the thermodynamic neutral ($\Delta G_H \approx 0$ eV), demonstrating an improved HER performance compared to the WO_3 (001). In a similar fashion, Zheng's group have developed ample of oxygen vacancies on the WO_3 nano sheets and observed a superior HER electrocatalytic activity in acidic medium (Zheng et al. 2017). These nano sheets show a Tafel slope of 38 mV dec with overpotential of 38 mV to generate 10 mA/cm^2 current density. Also the doping of hetero atoms develops more catalytically active sites in these WO_x by changing the electronic as well as the surface morphology. Additionally hetero structures get developed by this doping shows a remarkable improved HER performance. Dai's group has synthesized the NiO/Ni core shell like structures on multi walled carbon nanotubes (Gong et al. 2014). Because of the synergistic effect of the metal, metal oxide and nano carbon, the hybrid structure shows excellent HER activity close to that of platinum based electro catalysts.

4.3   Metal Chalcogenides

Primarily, the bio-inspired catalysts considered as an important material for scalable production of hydrogen. These are a particular class of bio enzymes that catalyze the protons and electrons present in the media to generate molecular hydrogen at lower potential (close to the thermodynamic potential). However it remained impossible to adopt these enzyme based devices for commercial hydrogen production (Burgess and Lowe 2002; Eady 1996; Eckenhoff et al. 2013; Hallenbeck and Benemann 2002; McPherson and Vincent 2014). These enzymes are so efficient to catalyze the HER (similar activity as Pt) in natural environment and lose their activity in both the strong acid and alkaline media. But inspired by the structure and function of these enzymes, researchers have explored transition metal (like Mo, W, Fe, Ni, Co etc.) chalcogenides as eminent electrocatalysts for HER.

In 1970s, the researchers have tried to explore bulk MoS_2 for HER, but unfortunately they did not get any fruitful result. Afterwards, these were considered as inefficient material for HER. This may be due to the poor electrical conductivity and existence of less active sites in bulk MoS_2 (MoS_2 are semiconductors). Therefore numerous scientific strategies have been developed to improve electrical conductivity as well as active sites of MoS_2 thereby improving their catalytic performances. This was mainly performed either by the "active site engineering" or by "electronic conductivity engineering". In many cases, both the strategies have been integrated to achieve better performance. The former one can be achieved mainly by (i) availing more number of catalytic active sites, (ii) enhancing the reactivity of these active sites and (iii) developing better electrical contact to the active sites. Whereas the latter has been performed either by (i) doping the MoS_2 lattice structure with suitable dopants or by (ii) compositing with conductive materials like graphite, carbon nanotubes and graphene. Later on, in 2005, Hinneman et al. have observed the resemblance of (1010) of Molybdenum edge (Mo-edge) in MoS_2 with the active sites of nitrogenase (Hinnemann et al. 2005). Additionally, the theoretical studies reveal that the free energy of hydrogen bonding to Mo-edge is very close to that of Pt. These findings claim the possibility to use MoS_2 as active electrocatalyst for HER. This was the first study, in which the edge structure of MoS_2 has been properly used. Out of different methodologies, the nanostructurization process has been considered as most favorable method to increasing the catalytic active sites in MoS_2 (Benck et al. 2014; Lukowski et al. 2013; Voiry et al. 2013a; D. Wang et al. 2014; L. Zhang et al. 2014). Further the composite of MoS_2 nanoparticle with graphite has been prepared and their HER performance was also agreed with the theoretical observations. Afterwards a variety of MoS_2 nanoparticles with variable dimensions were synthesized and observed the direct relationship of the catalytic activity of MoS_2 nanoparticle with the edge length rather than their area coverage (Jaramillo et al. 2007). Therefore various synthetic methodologies have been formulated to reduce the dimension of MoS_2 into different surface morphologies. As the MoS_2 has layered like structure possessing higher specific surface area, so it favors to form sheet like nanostructures. Therefore much more work has been carried out on incorporation of active sites onto these MoS_2 nano sheets.

Particularly the liquid phase exfoliation of MoS_2 sheets is widely employed to prepare single/few layered MoS_2 nanomaterials. It involves the intercalation of lithium compounds (i.e. methyl lithium, n-butyl lithium, tert-butyl lithium etc.) among the sheets of bulk MoS_2 followed by exfoliation on treating with water. It has been observed that during this chemical exfoliation process, transformation of thermodynamically favored 2H phase to metastable 1 T takes place. These are basically two different phases of MoS_2 where the former comprised of two S-Mo-S layers of edge shared MoS_6 trigonal prisms and latter have only one S-Mo-S layer derived from the edge shared MoS_6 octahedra. Out of these two phases, the 1 T possesses superior electrocatalytic activity towards HER, which may be due to both the higher electrical conductivity and existence of more number of active sites in 1 T phase (Lukowski et al. 2013; Voiry et al. 2013a). On the other hand, exfoliated MoS_2 prepared by using the n-butyl lithium and tert-butyl lithium shows better HER performances than the methyl lithium compounds (Ambrosi et al. 2015). Also the phase transformation is possible by electrochemically pretreating the electrodes in an optimized potential window that also supported by the theoretical calculations. Therefore in some cases repeated cyclization of MoS$_2$ modified electrodes are followed prior to data collection. Like exfoliation, the ball milling and ultra-sonication processes are widely adopted for synthesis of active layered MoS$_2$ nanostructures (Gopalakrishnan et al. 2014; D. Wang et al. 2013; Wu et al. 2013). In addition to the above discussed top down methods, solvothermal/hydrothermal like bottom up approaches were also adopted for successful synthesis of nanostructured MoS$_2$. By varying the starting precursors, reaction temperature and time, MoS$_2$ nanostructures with variable surface morphology has been developed (Chung et al. 2014; Lu et al. 2014b). Also the chemical vapor deposition technique produces highly pure, thickness controlled MoS$_2$ nanostructures uniformly grown upon the predefined substrates (Chung et al. 2014; Lu et al. 2014b).

In a particular work, Yu et al. have prepared a thin layer of MoS$_2$ on glassy carbon surface with a controlled film thickness (Yu et al. 2014). The authors have observed a~4.47 fold decreased exchange current density on addition of every single layer of MoS$_2$. It is known that the HER is a surface phenomenon and hoping of electrons to the surface by crossing the interlayer potential barrier takes place during the catalysis process (Fig. 4.4). Thus on increasing thickness, the electron hoping is affected thereby decreasing the HER performance. Therefore thickness optimization of the electrocatalyst film is strictly required to achieve better catalytic activity.

Another effective way to improve catalytic activity of MoS$_2$ is by increasing the surface area by creating pores in them (Chen et al. 2011; Kibsgaard et al. 2012; Lu et al. 2013; Yang et al. 2014). Not only it will maximize the number of active sites

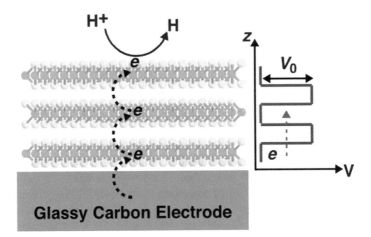

Fig. 4.4 Hopping of electrons in the vertical direction of MoS$_2$ layers. The right side illustrates the potential distribution in the multilayer film and the hopping of electrons through the potential barrier in the interlayer gap. (Reproduced with permission from Yu et al. 2014)

but also provide better contact of catalyst to the reactants and shorten the mass transport path thereby enhancing the catalytic performances. By using thiourea (as sulphur source) and Mo substrate, Lu et al. have prepared porous MoS_2 thin films (thickness: 400 nm–1.3 mm) following a hydrothermal synthesis method (Lu et al. 2013). An excellent HER performance has been observed with Tafel slope of 41–45 mV/dec and exchange current density of 2.5×10^{-7} mA/cm². Further doping of metallic elements (Co, Ni, V, Li etc.) into the lattice structure of MoS_2 significantly improve the electrocatalytic performances. For example, in case of Co or Ni doped MoS_2 nanostructures, the dopants (Co or Ni) preferably locate at the S-edge of MoS_2. This provides more number of exposed active sites thereby reducing the hydrogen adsorption free energy showing superior HER performance than undoped MoS_2 (Bonde et al. 2009; Lv et al. 2013; K. Zhang et al. 2014). However the V doping does not increase active sites but improves the electrical conductivity of MoS_2 thus increasing HER performances (Sun et al. 2014). In addition to these metal doping, incorporation of non-metallic atoms (such as O and N) into the crystal lattice of MoS_2 enhances the HER performance by increasing the electrical conductivity (Xie et al. 2013; W. Zhou et al. 2014). Like crystalline MoS_2, amorphous MoS_x also afford superior HER performance on doping with suitable dopants (Mn, Fe, Ni, Co, Cu, Zn etc.) (Merki et al. 2012). Here, the dopants are acting as the effective promoters and enhance the catalytic activity of MoSx nanostructures. The MoS_2 like electronic and structural properties of tungsten sulphides (WS_2) make them to draw substantial interest for HER. Therefore numerous synthetic protocols have been designed to prepare WS_2 with variable surface morphology (Choi et al. 2013; Pu et al. 2014a; Tran et al. 2013; Voiry et al. 2013b; Yang et al. 2013). Though the doping of foreign atoms provides better catalytic performance to the WS_2 nanostructures, but the actual region behind it is yet to explore (Tran et al. 2013). Afterwards, the electrocatalytic HER performances of VS_2, FeS, CoSx and NiSx are well studied (Di Giovanni et al. 2014; Faber et al. 2014b; Kong et al. 2013a). Further the participation of both the basal and edge site vanadium disulphide in the catalysis process makes it a potential candidate for HER. Thus many research group have synthesized vanadium sulphides of different surface morphology and explored their HER performances (Chen et al. 2017; Liang et al. 2016; Rao et al. 2017). Very recently, Behera's group has synthesized the VS_2 following a single step hydrothermal method and studied it's HER performances in 0.1 M H_2SO_4 electrolyte (Das et al. 2018). The VS_2 modified electrode initiates the catalysis process at very low onset potential (15 mV vs. RHE) and shows excellent cyclic durability maintaining the catalytic activity up to 43 hour of electrolysis. On the other hand, the FeS shows lower HER performances but their robust long term durability attracts the researchers to work on it. Therefore a variety of binary sulphides were derived by integrating these Fes (like NiFeS) for HER.

Since selenium (Se) has similar electronic configuration to sulphur and present in same group of chalcogenides in periodic table, the metal selenides gains substantial attention for electrocatalysis. Thus efforts have been paid and selenides of molybdenum ($MoSe_2$), tungsten (WSe_2), nickel ($NiSe_2$) and cobalt ($CoSe_2$) were synthesized. Contrary to sulphides, the (0001) basal planes of these selenides are not catalytically active but the activity of originates from the exposed active edge sites. It has been conformed from the DFT calculations that the Mo-edge in $MoSe_2$ and Se edge in $MoSe_2$ and WSe_2 played vital role in HER activity enhancement. Following bottom up synthesis approach, Zhou et al. have prepared a hierarchical $MoSe_{2-x}$ ($x = 0.47$) nano sheets from $MoO_2(acac)_2$ and dibenzyl diselenide (X. Zhou et al. 2014). The nano sheets were rich in Mo, which might provide higher electrical conductivity and more catalytically active sites for HER. Thereafter following similar synthetic strategies WSe_2, PbSe, SnSe nanostructures were developed. In addition to the powdery samples, supported metal selenides (hybrids) were also be synthesized using various conductive substrates (nickel foil, tungsten foil, carbon fiber etc.).

The strong interfacial interaction and additional conductivity make these supported hybrids to demonstrate superior HER performances along with better cyclic durability over the unsupported one. Moreover, the $MoSe_2$ and WSe_2 are having similar crystal structure like that of the MoS_2, where the individual sheets are stacked over one another (by means of van der Waals force of attraction) forming layered structure (Fig. 4.5). This type of configuration avails more exposed basal planes with negligible edge sites showing poor HER performances. Therefore, different synthesis processes were designed to prepare $MoSe_2$ nanostructures with more active edge sites. In a particular work, Cui et al. have prepared vertically aligned $MoSe_2$ film by selenizing the Mo film covered on an oxidized silicon substrate. The as obtained $MoSe_2$ surfaces completely covered with the edge sites thereby showing superior HER activity (Kong et al. 2013b). Similarly by using carbon fiber paper, the same group has grown vertically aligned $MoSe_2$ and WSe_2 with maximum amount of edge sites (H. Wang et al. 2013).

Though the transition metal chalcogenides are very active for HER, but some reports also available on their OER performances. Particularly the cobalt and nickel sulphides are active towards OER and iron sulphides shows poor performance. However the bimetallic sulphides formed by Fe with CoS_2 (Co-Fe-S) shows excellent OER catalytic activities (Shen et al. 2015). By tuning the synthesis procedures, various nickel sulphides with different surface morphologies are prepared and explored their superior OER activities (Chung et al. 2015; Feng et al. 2015; N. Jiang et al. 2014). The presence of continuous network like Ni-Ni bonds in Ni_3S_2 lattice make them to behave like metals (Ensafi et al. 2016; Ouyang et al. 2015; Zhou et al. 2013, 2015). Therefore among other nickel sulphides, the Ni_3S_2 shows superior OER performances. Other nickel sulphides are OER active, but show poor cyclic stability in harsh OER electrolytic conditions. On the other hand, it has been found that most of the cobalt sulphides are not active for OER and a few reports are available on it. To improve the catalytic behaviour, the cobalt sulphides are allowed to grow onto metal oxide layer (Jing Yang et al. 2016). Here the metal-oxygen linkage the cobalt sulphide surface helps to break the Co-O bond and form Co-O-O superoxide that participates in the catalytic process. Liu et al. have synthesized cobalt sulphide by electro deposition method and observed better OER activity (Liu et al. 2015).

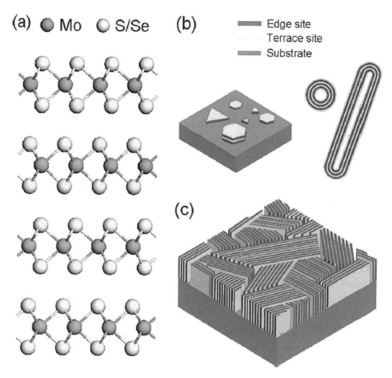

Fig. 4.5 Nanostructures of layered MoS_2 and $MoSe_2$. (**a**) Layered crystal structure of molybdenum chalcogenide with individual S–Mo–S (or Se–Mo–Se) layers stacked along the c-axis by weak van der Waals interaction. The highly anisotropic crystal structure is the origin of anisotropic electrical and chemical properties. (**b**) Schematics of MoS_2 nanoparticles with platelet-like morphology distributed on a substrate (left), and nanotubes and fullerene-like nanotubes of MoS_2 and $MoSe_2$ (right). (**c**) Idealized structure of edge-terminated molybdenum chalcogenide films with the layers aligned perpendicular to the substrate, maximally exposing the edges of the layers. (Reproduced with permission from Kong et al. 2013b)

The as deposited electrode requires only 361 mV overpotential to deliver the state of the art current density (10 mA/cm²) and better cyclic stability. Two forms of cobalt selenide i.e. CoSe and $CoSe_2$ are active for both the OER, HER and ORR (Feng et al. 2008; Kong et al. 2014; Y. Liu et al. 2014). However, the bulk $CoSe_2$ shows poor activity, thus they are subjected to exfoliate to form nanosheet like structures availing active Co-sites. The presence of active sites and vacancies attribute this improved OER performance (Y. Liu et al. 2014). As discussed above, a number of works on the synthesis and catalytic performance study of transition metal chalcogenides were carried out, but their performances are still behind that of noble metal nanostructures. Therefore more work is needed to (i) design innovative synthetic methods for scalable preparation of TMC, (ii) incorporate more active sites and (iii) improve their catalytic performance as well as cyclic stability to reduce the reaction overpotential.

4.4 Metal Phosphides

The transition metal phosphides (TMP) are considered as promising electroactive materials to replace noble metal based catalysts for the electrochemical water splitting (both for HER and OER). Inspired by the catalytic activity of [NiFe] hydrogenases towards biological path ways, Liu and Rodriguez have observed an excellent activity of Ni_2P (001) for HER over the platinum group electrocatalysts (Liu and Rodriguez 2005). Basing upon the density functional theory (DFT) calculations, the phosphorous (P) atom in these TMPs plays a crucial role in enhancement of the catalytic activity. The synergistic effect of proton acceptor P sites and hydride acceptor metal site on the surface of TMPs promotes the evolution of hydrogen. Here the phosphorous atoms possess stronger electronegativity making withdrawal of the electrons from the metal surface and traps the positively charged protons effectively during the HER process (Liu and Rodriguez 2005; Xiao et al. 2014). It has been observed from the previous works that the increasing P concentration improves the HER activity to a certain extent (Callejas et al. 2015; Pan et al. 2015). Further addition restricted the delocalization of electrons around metal atoms of TMPs showing a decreased conductivity thereby lowering the HER performances (Blanchard et al. 2008; X. Wang et al. 2015). Therefore an appropriate ratio of P to metal ion should be followed for better utilization of catalytic active sites of the TMPs. Initially, high temperature/pressure methods were used for the synthesis of TMPs with flammable elemental phosphorous or noxious phosphine gas as the source of phosphorous restricting the scalable synthesis and application of TMPs. Afterwards various synthetic protocols have been formulated that uses variety of phosphorous precursors including organic (i.e. trioctylphosphine), inorganic (i.e. hypophosphite, phosphate) and elemental phosphorous (i.e. red phosphorous) (Pan et al. 2015; Muthuswamy et al. 2009; P. Jiang et al. 2014a; Yang et al. 2015; Xing et al. 2014; Pu et al. 2014b; Wu et al. 2017a). In general, the synthesis process of these TMPs strictly depends on the source of phosphorous precursor chosen. For example, organic phosphorous are taken in case of wet chemical synthesis methods whereas the high temperature solid phase synthesis method use the elemental phosphorous as the source phosphorous. On this basis, the synthesis of these TMPs can be divided into the following three categories,

4.4.1 Route-(I) Use of Organic Phosphorous

By using organic phosphorous (mainly trioctylphosphine) as the phosphorous source, a variety of TMPs were derived following the wet chemical synthesis methods. Generally, the C-P linkage can be easily broken at temperature >320 °C, therefore solvents having higher boiling points like oleylamine, octadecene and squalene etc. are used to assure the formation of metal phosphides. In this regard highly dispersive TMPs like CoP, Ni_2P and FeP nanoparticles were derived from the metal acetylacetonate (as metal precursor) and trioctylphosphine (as phosphorous source) following this wet chemical method (Callejas et al. 2014; Popczun et al. 2013, 2014).

$$\left(C_8H_{17}\right)_3 P + M\left(acac\right)_2 \rightarrow MP_x \qquad (4.1)$$

Here, the phase, surface morphology and size of these metal phosphides can be manipulated by tuning the reaction conditions (like type of solvent taken, reaction temperature and ratio of phosphorous to metal) (Pan et al. 2015; Seo et al. 2016). Since during the reaction, highly corrosive and flammable phosphorous are released, therefore it's recommended to take precaution and to carry out the whole reaction only in inert atmosphere.

4.4.2 Route-(II) Use of Inorganic Phosphorous

Instead of the noxious organic phosphorous, the inorganic phosphorous like hypophosphite (i.e. $NH_4H_2PO_2$, NaH_2PO_2) and phosphates ($(NH_4)_2HPO_4$) were used as the source of phosphorous for TMP synthesis. In these cases, the in situ formed PH_3 gas treated with the as formed metal oxides/hydroxides, metal organic frameworks etc. and generate the corresponding phosphides as per the following equation,

$$2NaH_2PO_2 \rightarrow Na_2HPO_4 + PH_3 \uparrow \qquad (4.2)$$

$$PH_3 + M_xO_y \rightarrow MP_z \qquad (4.3)$$

The use of inorganic phosphates maximizes the retention of precursor morphology and dimension. For example, FeP nanowires, WP_2 sub micro particles and CoP/CNT nano hybrids were synthesized through the low temperature hypophosphate based reactions (P. Jiang et al. 2014b; Q. Liu et al. 2014; Xing et al. 2015). On the other hand, the molybdenum phosphides, tungsten phosphides etc. are synthesized taking the phosphates as phosphorous source via a hydrothermal, calcination and annealing sequentially (Wu et al. 2017a; Xiao et al. 2014; Xing et al. 2015, 2014).

4.4.3 Route-(III) Use of Elemental Phosphorous

The elemental phosphorous is generally used as the phosphorous source for the preparation of phosphorous-rich metal phosphides via chemical vapor deposition method. This has been achieved by tuning the metal to phosphorous ratio as well as the reaction temperature (Jiang et al. 2015; Laursen et al. 2018; Pu et al. 2014b).

For example, phosphorous enriched FeP_2/C nano hybrids and WP_2 nano wire arrays were developed through solid phase reaction taking the red phosphorous as the elemental phosphorous source (Jiang et al. 2015; Pi et al. 2016).

$$P + M \rightarrow MP_x \qquad (4.4)$$

The resemblance with [NiFe] hydrogenase, motivated the researchers to explore nickel metal based phosphide (NiP_x) catalysts for HER. It has been observed that on changing the atomic ratios of nickel and phosphorous, nickel phosphide with different crystal phases was prepared showing a variable electronic as well as physicchemical behaviour. Till now numerous work in the designing of both the metal rich (x < 1, i.e. Ni_2P, Ni_5P_4, Ni_3P, $Ni_{12}P_5$) and phosphorous rich (x > 1, i.e. NiP_2) phases of NiP_x nanostructures were developed and their application towards electrochemical HER has been explored. In a particular work, Sun et al. have developed NiP_2 nano sheets (NSs) on carbon cloth (as supporting material) following route-II (as discussed above) using hypophosphite as phosphorous precursor and studied the HER activity in 0.5 M H_2SO_4 (P. Jiang et al. 2014c). These NSs showed excellent HER performance requiring only 74 mV of overpotential to deliver 10 mA/cm^2 (η_{10}) current density with better cyclic durability (up to 57 hours). In another report, Dismukes et al. have synthesized highly conductive Ni_5P_4 micro particles (size is around 0.3 ~ 1.8 μm) and demonstrated electrochemical evolution of hydrogen in both the acidic and alkaline medium (Laursen et al. 2015). Interestingly, HER activity close to the Pt based electrocatalyst has been noticed in 1.0 M H_2SO_4 electrolyte. It needs only 23 mV overpotential to deliver 10 mA/cm^2 with a lower Tafel slope of 33 mV/dec. Also in alkaline medium (1.0 M NaOH), a remarkable catalytic performance has been observed with η_{10} of only 49 mV. But a restricted HER performance has been observed with further reducing the concentration of phosphorous. By using organic phosphorous precursors, Schaak's group has synthesized hollow and multifaceted Ni_2P nanoparticles (Popczun et al. 2013). These nanostructures exhibited excellent HER performances requiring only 20 mV of overpotential to achieve 20 mA/cm^2. In addition to the single phase NiP_x nanostructures, biphasic Ni_5P_4-Ni_2P nano sheet arrays also developed (X. Wang et al. 2015). These nano sheets shows better HER performance in terms of lower onset potential with robust cyclic durability in acidic medium.

Based upon the previous reports, the metallic cobalt has been theoretically calculated to have lower energy barrier for the hydrogen adsorption, therefore the CoP_x are assumed to be the promising electrocatalysts for HER (Lin et al. 2011; Zheng et al. 2015). The cobalt based phosphide electrocatalysts (CoP_x) are the important member of transition metal phosphides demonstrating higher HER activity with resistance to acidic medium. Among them the CoP electrocatalysts are extensively studied for HER. They possess MnP like crystal structure in which the phosphorous atom is surrounded by six cobalt atoms forming a highly distorted triangular prism like structure (Popczun et al. 2014). Hence CoP nanocrystals with different surface morphologies like nano particles, nano sheets, nano wires, urchin like etc. has been designed by tuning the reaction conditions as well as the starting precursors

(Popczun et al. 2014; Tian et al. 2014a; Yang et al. 2015; Zhang et al. 2017). For the first time, Schaak's group has taken the initiative and successfully synthesized the CoP nano particles (particle size of 13 ± 2 nm) following Route-(I) with the organic phosphorous sources (Popczun et al. 2014). These nano particles on Ti foil shows better HER activity generating 20 mA/cm^2 at only 85 mV of overpotential in 0.5 M H$_2$SO$_4$ electrolyte. Thereafter an enhanced electrocatalytic activity of one dimensional CoP nano wires has been recorded in all pH electrolytes (ranging from 0 to 14). Furthermore taking the benefits of large surface area, suitable structural distortion, and higher proton/electron transfer efficiency and ample of reactive active sites, Zhang et al. have developed two dimensional CoP nano sheets (Zhang et al. 2017). These nano sheets were synthesized by phosphidising the Co$_3$O$_4$ precursors following Route-(II) of the TMP reaction processes. The highly active (200) crystal face of CoP NSs shows better HER performances with 131 mV overpotential at 100 mA/cm^2, lower Tafel slope (44 mV/dec) and durability up to 24 hours. Also by varying the ratio of Co: P and reaction conditions, different cobalt phosphide like Co$_2$P, CoP$_2$, and CoP$_3$ etc. nano crystals were synthesized and their HER activity has been explored. In a particular work, Wu et al. have synthesized CoP$_3$ concave polyhedron like structure by phosphidising the MOF derived Co$_3$O$_4$ (Route-(II) of TMP reaction process) (Wu et al. 2017b). These polyhedrons show a remarkable HER performance with overpotential of 78 mV at 10 mA/cm^2 in 0.5 M H$_2$SO$_4$ and durability in all pH electrolytes.

Because of the similarity in active sites of [FeFe] hydrogenase and abundant reserve of iron, the iron phosphide was thought to be the active low cost material for industrial HER application. Various iron phosphides have been developed through different reaction methodologies but FeP, Fe$_2$P and Fe$_3$P shows superior electrocatalytic HER performance. Jiang et al. have synthesized FeP nano wires that catalyzes the HER at a lower onset potential (16 mV) and need only 55 mV for η_{10} in 0.5 M H$_2$SO$_4$. Among these three phosphides, the Fe$_3$P perform better and needs only 49 mV overpotential to deliver 10 mA/cm^2 in acidic medium (Schipper et al. 2018). This experimental observation is also supported by DFT calculations stating that the hydrogen prefers to bind to the Ferric region availing higher hydrogen coverage on iron enriched FeP$_x$ surfaces. But these FeP$_x$ electrocatalysts shows poor durability owing to the corrosion taking place on its surface. Therefore carbon shell supported FeP nanoparticles were developed following both the phosphidation and carbonization processes (Fig. 4.6a) (Chung et al. 2017). During the carbonization process, the polymer coating converted to the carbon shell and Fe$_2$O$_3$ converts to FeP at the time of phosphidation reaction. The carbon shell protects the FeP present in core from oxidative corrosion and shows better HER performance as well as durability in acidic medium (Fig. 4.6b, c). The resistive behaviour of the core-shell structure has been verified by the EXAFS (Extended X-ray absorption fine structure) analysis. As can be seen from the following figure (Fig. 4.6d–e), the bare FeP shows a slight shifting towards left after 5000 repeated cycles, while a little difference has been realized in the core-shell structure.

Likewise, the copper based metal phosphides were developed and either HER performances were broadly studied. The Cu$_3$P considered as the potential candidate

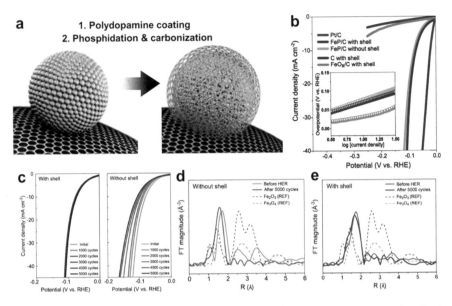

Fig. 4.6 (**a**) Schematic representation of carbon-shell-coated FeP nanoparticles preparation. (**b–c**) Polarization curves of FeP with and without a carbon shell in 0.5 M H$_2$SO$_4$. The corresponding Tafel plots are shown in the inset of (**b**). (**d–e**) Extended X-ray absorption fine structure (EXAFS) analysis of FeP nanoparticles without and with the carbon shell. (Reproduced with permission from Chung et al. 2017)

for HER. In this regard, many researchers have designed the synthetic protocols and derive the Cu$_3$P based nano wires, nano sheets and nano particles. Sun's group has synthesized the Cu$_3$P nano wires (NWs) directly on copper foam surface by following the hypophosphite based reaction route (Tian et al. 2014b). In acidic medium (0.5 M H$_2$SO$_4$), these Cu$_3$P modified electrode catalyze the HER efficiently at η_{10} = 143 mV with Tafel slope of 67 mV/dec. In comparison to the nano wires, the Cu$_3$P nano sheets performs well towards HER showing η_{10} of 105 mV with Tafel slope of 42 mV/dec in 1.0 M KOH electrolyte. But the HER activity of the CuP$_x$ nanostructures still remain far apart from the Ni, Co and Fe based electrocatalysts. Therefore, efforts have been paid to improve the catalytic efficiency and core shell like structures with carbon has been developed. Wang et al. have synthesized the hierarchical Cu$_3$P nanostructures in N, P-codoped carbon shell (Cu$_3$P@NPPC) and noticed an excellent HER performance (η_{10} = 89 mV) (Wang et al. 2018). Inspired by the superior hydrodesulphurization activity, the molybdenum phosphides are presumed to be an efficient electrocatalyst for HER. Since the electronic structure of MoP$_x$ is similar to that of platinum, so these MoP$_x$ nano structures are considered as the potential TMPs for HER. Following phosphate based reaction route, Xiang et al. have prepared MoP nanostructures and observed the role of extent of phosphorization in HER performance enhancement. The DFT studies revealed that the MoP systems required nearly zero binding energy for hydrogen atoms, thereby the P atoms adsorbs hydrogen at a low coverage and desorbs/delivers hydrogen at higher coverage. Therefore these MoP based HER electrocatalysts performs well both in the acid and alkaline medium. In acidic medium (0.5 M H_2SO_4), it requires only 184 mV of overpotential to deliver 30 mA/cm^2 with a Tafel slope of 54 mV/dec. However the high temperature phosphorization process lead to formation of agglomerated and sintered MoP nanostructures showing reduced HER activity. Therefore to achieve the actual electrocatalytic performance of MoP, hetero structures with carbon nano materials (i.e. carbon nano tubes, porous carbon, graphene etc.) were developed. The herero structure shows an uniform distribution of small sized MoP nanoparticles throughout the CNT surface (X. Zhang et al. 2018). This MoP/CNT shows excellent electrocatalytic HER activity and durability in all pH condition (both in acidic, neutral and alkaline condition). Not only CNTs but also the composite formation with porous carbon material add additional HER activity and durability to the MoP nanostructures (Jian Yang et al. 2016). Thereafter an improved electrocatalytic activity of MoP has been achieved by compositing with the porous carbon (MoP@PC). In addition to the molybdenum based TMPs, the tungsten phosphides are also explored in form of WP, WP_2 etc. Further by tuning the reaction condition and employing different supporting substrate (Ti foil, carbon cloth,), the WP amorphous nanoparticles, 2D WP nano rods on carbon cloth, WP nanoparticles in heteroatom doped carbon matrix, WP_2 sub micro structures etc. and explored their HER performances (McEnaney et al. 2014; Pu et al. 2016, 2014b; Xing et al. 2015).

As discussed in the above section, the TMPs shows superior HER activity and the composites with carbon nanostructure further provides an additional durability, but their performance still remains far apart from noble metal nanostructures. Since both the DFT calculations and experimental observations precluded that the electronic and crystal structure of these metal phosphides plays an important role in tuning their catalytic activity. Therefore efforts have been paid to develop new TMPs by substituting either cation (i.e. metal) or anion (i.e. P) or both in the metal phosphides. Here the composition of these binary metals regulates the electronic structure promoting the intrinsic electrocatalytic activity towards HER. In a report, Jaramillo et al. have studied the variation in hydrogen adsorption free energy, HER reduction current density and turn over frequency by varying the ratio of Fe and Co in mixed metal phosphides (Kibsgaard et al. 2015). Among them the $Fe_{0.5}Co_{0.5}P$ shows an optimum ΔG_{H*} with higher reduction current density and average TOF. Later on an improved activity of the binary metal phosphide (BTMPs) was achieved by compositing or by growing on the substrates like carbon cloth. By varying the concentration of Fe, a series of BTMPs of $Fe_xCo_{1-x}P$ (0 < x < 1) nanowires on carbon cloth developed and their HER activity has been demonstrated (C. Tang et al. 2016). Among them the $Fe_{0.5}Co_{0.5}P$ shows superior HER activity (similar to Pt) requiring only η_{10} of 37 mV with excellent durability (activity remained stable up to 100 hours). Also the substitution of phosphorous atom with sulphur atoms in the TMPs to form the transition metal phosphosulphides (TMPSs) considered to an effective method to tune the HER activity of TMPs. Because of the difference in electronegativity values, the introduction of sulphur shifts the d-band center in these TMPs and alters the HER activity. In this regard, TMPSs like MoPS, CoPS etc. and their composites with carbon cloth and carbon nanotubes were developed and observed the HER activity close to that of the Pt based electrocatalysts (W. Liu et al. 2016; Ye et al. 2016).

Inspired by the superior electrocatalytic HER performance of the TMPs, their OER activity also studied. In this regard, many phosphide based catalysts like CoP, Co_2P, Ni_2P, Ni_3P_5, FeP etc. were developed following various synthetic methods and their anodic performance towards OER been explored in alkaline medium. In these cases, a significant catalytic activity has been noticed, but the post catalysis characterizations reveal the chemical transformation of the catalyst surface. During the course of reaction the surface of this phosphide get oxidized forming corresponding metal oxide/hydroxides and oxyhydroxides, therefore instead of catalyst they are named as "pre-catalyst". That's why, till now no such theoretical study has been carried out on the OER catalytic activity of these TMPs. Despite of the surface transformations, the TMPs shows remarkable catalytic activity compared to some of the leading catalysts. As like in case of their HER activity, the OER performance also can be tuned by introducing both cationic and anionic impurities. In a particular work, Brock's group have incorporated Mn atoms into the Co_2P lattice and observed a significant reduction in reaction overpotential (D. Li et al. 2016). Not only the Mn, but also the introduction of Fe atoms into the Co_2P lattice improves the catalytic activity (Mendoza-Garcia et al. 2015). The as formed ternary metal phosphide ($Co_{(2-x)}Fe_xP$) shows superior activity in comparison to both the Co_2P and Fe_2P. Likewise the Ni doped CoP also shows better catalytic performance under similar experimental conditions (Y. Li et al. 2016). Also the insertion of anions into the metal phosphide increases the OER activity. Duan et al. stated in their report that not only the cations, but also the incorporation of anions improves the OER activity of metal phosphides (Duan et al. 2016). In their work they have prepared oxygen doped and oxygen, iron dual doped Co_2P and observed better performance compared to that of the undoped one. The presence of additional cation and anion provides extra electrical conductivity to the corresponding metal phosphides demonstrating reduced charge transfer resistance facilitating the charge transfer rate. Additionally the doping provides more catalytic active sites thereby reducing the reaction overpotential thus showing improved catalytic activity.

Since the catalytic process is a surface phenomenon, so activation of the catalyst surface is recommended in these TMPs. With a close look, one can find an irreversible per oxidation peak prior to the onset potential during the first voltammetric sweep of almost all the metal phosphides (Dutta et al. 2016). On increasing the sweep number this peak disappeared lowering the reaction overpotential as well as Tafel slope. This may be the catalyst activation that leads to the possible chemical transformation of the catalyst surface (Dutta et al. 2016; Ryu et al. 2015). Dutta and Samantara et al. have synthesized needle shaped narrow hexagonal phase one dimensional Co_2P nanostructure following a phosphine gas mediated synthesis method and studied the HER as well as OER performances (Fig. 4.7) (Dutta et al. 2016).

The catalyst surface has been activated by five cyclic voltammetric scans (in range of 1.20–1.65 V vs. RHE) prior to record the linear sweep voltammograms (LSVs). In 1 M KOH, the catalyst required only 310 mV overpotential to deliver

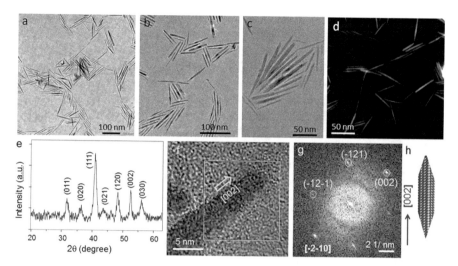

Fig. 4.7 (**a–c**) TEM images of needle-shaped Co$_2$P nanostructures at different resolutions. (**d**) HAADF-STEM image of Co$_2$P nanostructures. (**e**) Powder XRD pattern of Co$_2$P nanostructures. (**f**) HRTEM image of a tapered nanorod. (**g**) Selected area FFT pattern. Planes are labeled as per the viewing direction. (**h**) An atomic model of a typical 1D nanostructure of Co$_2$P. (Reproduced with permission from Dutta et al. 2016)

10 mA/cm^2 current density with a Tafel slope of 50 mV/dec and better cyclic durability up to 10 hour. Here the surface activation leads the transformation of cobalt phosphide pre-catalyst to the surface oxidized catalyst thereby showing superior catalytic activity. In another report the same group has developed a core-shell structured nickel phosphide with Fe$_2$O$_3$ core and amorphous Ni$_x$P and observed excellent OER performances (Dutta et al. 2018). Though ample of work in the field of metal phosphides has been carried out, but still the actual mechanism behind the catalytic activity of the pre-catalyst is unclear. Therefore more theoretical as well as experimental works should be carried out and the complete mechanism should be drawn for better understanding.

4.5 Supported and Free Standing Catalysts

As discussed in the above section, the electrocatalysts are coated onto the conductive electrode surface and dried properly prior to the electrolysis (HER/OER). This process is widely used in case of the powdered catalyst samples. Because of continuous gas evolution, sometimes the coated catalyst observed to be peeled off from the electrode surface impairing the cyclic durability and catalytic performance. Therefore the self-supported electrocatalysts either by direct growing on conductive substrates or making free standing films are developed following different synthetic protocols. Generally by using the carbon cloth, metal foil (Ni, Cu, Ti foils) and 3D metallic substrates (Ni foam, stainless steel mesh etc.), the self-supported electrocatalysts are synthesized following hydrothermal/solvothermal, chemical/ physical vapor deposition (CVD/PVD) and electro deposition methods. Further on the basis of the mechanical strength and surface structure, the conductive substrates are categorized as (i) rigid substrates, (ii) soft substrates and free standing catalyst films. These single step facile synthetic procedures does not require any organic binders and post synthesis treatments demonstrating better contact with the electrode surface and well integrity of the catalyst films. The good electrical conductivity, mechanical stability and availability of more catalytically active surface area make these self-supported/free standing catalysts for better exposure and enhanced utilization of electroactive sites as well as shows excellent operational durability compared to the coated catalysts. Moreover the complex structure and three dimensional arrays of the nanostructured catalysts on the conductive substrate facilitate the ion diffusion process and movement of electrolyte facilitating the evolution of the reaction products (H_2/O_2). Therefore the synthesis of these self-supported/free standing electrodes through a suitable synthesis process assumed to have a potential for direct application in the practical energy conversion devices.

Initially the powder samples were prepared by different synthesis processes followed by coating onto the electrode surface to study the catalytic performances. In these cases, the continuously evolved oxygen bubbles during the OER process blocks the flow of ions /electrons to the active sites leading to a significant performance loss of the catalyst. Also peeling off of the coated catalyst layer from electrode surface limits their cyclic durability (Fig. 4.8). Additionally, the coated powder samples form irregular aggregated voids on planar electrode surfaces making unfavorable gas transport demonstrating poorer reaction kinetics. Therefore, development of throughput gas breathing electrocatalyst became highly indispensible for smooth running of the electrolyzer. In this regard, effort have been paid and self-supported nanostructures were developed on various metal surfaces like Pt, Au, Co, Pd and Cu etc. that performs many fold better than that of the unsupported nanostructures as well as the state of the art catalysts (Ir) for OER (Lee et al. 2012; Over 2012). The electronegativity of the support metals in these supported nanostructures plays a great role in monitoring their electrocatalytic performances as in the following sequence Au/CoO$_x$ > Pt/CoO$_x$ > Pd/CoO$_x$ > Cu/CoO$_x$ > Co/CoO$_x$.

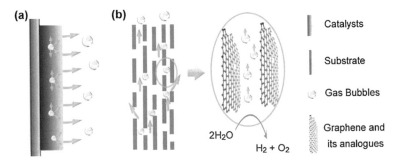

Fig. 4.8 Schematic gas evolution in (**a**) powdery catalysts coated on planar substrates like glassy carbon, and (**b**) free-standing graphene-based films, (Reproduces from Ma et al. 2016)

Since electrolysis is a surface phenomenon and the efficiency of the catalyst strongly depends on the accessible surface area and exposed active centers, so various nanostructures like nano flakes, wires, rods, porous spheres etc. are prepared both in powdery and self-supported form. Because of the short ion/mass transport path and more roughness factor and active sites, the one dimensional, porous nano wires and nano rods shows enhanced catalytic performances. Samantara et al. have prepared one dimensional porous $NiCo_2O_4$ nano rods on reduced graphene oxide (RG/NCO NRs) by hydrothermal process following a post annealing method. The RG/NCO NRs shows better electrocatalytic performance towards OER requiring only 330 mV of overpotential to achieve the benchmark current density (10 mA/ cm^2) (Samantara et al. 2018). Further, a smooth mass transfer, higher catalyst loading and better contact with the conductive surface has been achieved by incorporating the metal oxide nanostructures into three dimensional (3D) scaffolds like Ni-foams, stain less steel meshes, carbon cloth and carbon fibers. In a particular work, Liang et al. have deposited the Co_3O_4/N-doped graphene composite onto the Ni-foam with a loading up to 1 mg/cm^2 which is much higher than that of the powder sample coated onto the glassy carbon electrode (0.1 mg /cm^2). These self-supported electrodes observed to catalyze the OER efficiently in comparison to the powdery sample and needs only 310 mV overpotential to deliver the state of the art current density of (10 mA/cm^2) with very low Tafel slope of 67 mV/decade (Esswein et al. 2009; Liang et al. 2011). In another work, nanowires of Co/Ni based mixed oxide free standing electrodes on Ni-foam were prepared following hydrothermal process and observed the impact of Ni content on the nanostructure morphology (Lu et al. 2011). On increasing Ni-content, the nano flakes are obtained instead of nano wires, but the nanowires formed at Ni/Co concentration of 1:1 ratio perform better OER activity. By following the same hydrothermal method, Lee et al. have prepared the Co_3O_4 nanowires on stainless steel mesh. The as formed hierarchically mesoporous nanowires perform well as the bifunctional electrocatalyst for OER and ORR in alkaline medium (Lee et al. 2014). Here, the stainless steel mess not only acts as the support but also plays the role of current collector avoiding the use of binder, thereby significantly reducing the interfacial resistance facilitating its use in metal air batteries. In addition to these metal foils and foam/meshes, graphite plates and glass sheets are also used as the rigid substrates for the development of different shaped metal based nanostructures (Faber et al. 2014a; P. Jiang et al. 2014b). Though these rigid substrates have many advantages, but their heavy weight, lack of flexibility and bulky structure restricts the catalyst synthesis by the reactor (autoclaves, tube furnaces etc.) dimension.

Therefore carbon fiber paper (CFP) constituting the conductive micron level carbon fibers (7–10 μm) are broadly used. The flexibility, eco friendliness, light weight, better conductivity as well as good mechanical strength promotes these CFPs for the scalable preparation of supported electrocatalysts. Significantly the CFP provides a flexible/durable matrix for the growth of nanostructures, avails an expanded surface area showing more number of electroactive sites leading to better electron/charge transfer facilitating the catalysis process. Additionally, the miniaturization of these self-supported catalysts make suitable for their application in designing of foldable, wearable and portable energy systems (Liao et al. 2013; Tian et al. 2014a). These papers are commercially available or can be prepared by carbonizing the cellulose fiber paper. Prior to the synthesis of catalyst, the hydrophobic surface of the CFPs is converted to hydrophilic by means of surface oxidation through mild oxidation (treating with acids) or oxygen plasma (Gong et al. 2013; Kong et al. 2014). In a particular work, Ma et al. have prepared phosphorous doped graphitic carbon nitride (PCN-CFP) on the CFP surface and studied the catalytic performance for OER. The cleaned CFP was subjected to surface oxidation in which the oxygen functionalities like –COO⁻ get developed on its surface making it hydrophilic nature. By this process, large sized (10–15 cm) self-supported N and P co-doped catalysts were produced showing superior catalytic behaviour for both the ORR and OER (Fig. 4.9a, b). They have demonstrated the flexibility of the electrode by observing retention of OER current about 94.6% even after folded for three times (Fig. 4.9c).

Fig. 4.9 (**a, b**) SEM images and (inset in panel a) photograph of P-doped g-C3N4 growing on CFP (PCN-CFP). (**c**) Polarization curves of PCN–CFP in different folded and rolled-up forms, and (inset in panel **c**) variations of ORR (at 0.40 V) and OER (at 1.63 V) currents. (Reproduced from Ma et al. 2015)

Without using any additional conductive substrate, a different type of free standing electrode was designed for graphene films and its analogous 2D materials (like MoS_2, WS_2, g-C_3N_4 etc.). The ordered arrangement of active catalyst layers avails oriented channels with interconnected porosities facilitating the ion/electron transport as well as the gas evolution (Fig. 4.8b). Generally, these types of electrodes were obtained by vacuum filtration of the active catalyst dispersion where the thickness of the electrode can be controlled by changing the concentration and volume of the catalyst dispersion. Particularly, in case of graphene, the π-π interaction (among the graphitic domain) and electrostatic repulsion force (between the oxygen functionalities) makes equilibrium in the film. After vacuum suction, the freestanding electrode can be collected either by peeling off or by dissolving the filter membranes (Gilje et al. 2008). Along with graphene, various non-metal nanostructures were integrated in these film electrodes following the layer by layer assembly of the intended active catalyst dispersions.

Chen et al. have prepared a layer by layer assembly of N, O-dual doped graphene and carbon nanotubes and observed an enhanced catalytic performance towards OER (Chen et al. 2014). Both the active sites developed by doping of N and O provides ample of active sites, thereby requiring only of 368 mV of overpotential to deliver 5 mA/cm^2 current density. The catalytic behaviour of this film electrode observed to be better than the powder sample, dry NG-CNT film, undoped graphene-CNT film and also than that of the IrO_2 powder catalysts coated on conductive electrode surfaces. Various metal clusters, metal oxide/hydroxides etc. are synthesized onto these prepared free standing graphene films following electrochemical or chemical deposition methods. The oxygen functionalities (epoxy, hydroxyl, carboxylic etc.) present on reduced graphene oxide film acts as the anchoring sites for these nanostructures facilitating their well dispersion (Gilje et al. 2008). Chen et al. have prepared well dispersed Ni nanoparticles onto nitrogen doped graphene film through heterogeneous reaction (Chen et al. 2013b). The nitrogen centers present in doped graphene film facilitate the interaction of the graphene sheet with the metal precursors forming a uniform distribution of the metal clusters. Furthermore the higher electronegativity of nitrogen (3.0 eV) makes a strong interaction with the metal centers forming an M-N-C bond alleviating the electrochemical processes. The Ni NPs (66 nm in dimension) are decorated on both the sheet edges and base plane graphene sheet with a lamellar structure. These hybrid film acts as an efficient free standing electrocatalyst and catalyzed the OER with a comparable activity to that of the state of the art IrO_2. It has been observed that, electro deposition in alkaline condition (i.e. in presence of urea) form the metal hydroxide nanostructures onto the conductive graphene film that acts as an efficient electrocatalyst for water oxidation (Chen et al. 2013a). Further by annealing these hydroxide nanostructures, metal oxide based composite material are obtained. For example, Chen et al. have prepared hierarchical, porous $NiCo_2O_4$ nanostructures on paper like nitrogen doped graphene films and studied its OER performance (Chen and Qiao 2013). The large surface area and pore volume as well as better contact of the binary oxide with the graphene film afforded abundant electroactive sites facilitating the oxygen evolution.

References

Ambrosi, A., Sofer, Z., & Pumera, M. (2015). Lithium intercalation compound dramatically influences the electrochemical properties of exfoliated MoS2. *Small, 11*, 605–612.

Bau, J. A., Li, P., Marenco, A. J., Trudel, S., Olsen, B. C., Luber, E. J., & Buriak, J. M. (2014). Nickel/iron oxide nanocrystals with a nonequilibrium phase: Controlling size, shape, and composition. *Chemistry of Materials, 26*, 4796–4804.

Benck, J. D., Hellstern, T. R., Kibsgaard, J., Chakthranont, P., & Jaramillo, T. F. (2014). Catalyzing the Hydrogen Evolution Reaction (HER) with molybdenum sulfide nanomaterials. *ACS Catalysis, 4*, 3957–3971.

Bergmann, A., Martinez-Moreno, E., Teschner, D., Chernev, P., Gliech, M., de Araújo, J. F., Reier, T., Dau, H., & Strasser, P. (2015). Reversible amorphization and the catalytically active state of crystalline Co3O4 during oxygen evolution. *Nature Communications, 6*, 8625.

Blanchard, P. E. R., Grosvenor, A. P., Cavell, R. G., & Mar, A. (2008). X-ray photoelectron and absorption spectroscopy of metal-rich phosphides M2P and M3P (M = Cr-Ni). *Chemistry of Materials, 20*, 7081–7088.

Bonde, J., Moses, P. G., Jaramillo, T. F., Nørskov, J. K., & Chorkendorff, I. (2009). Hydrogen evolution on nano-particulate transition metal sulfides. *Faraday Discussions, 140*, 219–231.

Boppana, V. B. R., & Jiao, F. (2011). Nanostructured MnO2: An efficient and robust water oxidation catalyst. *Chemical Communications, 47*, 8973–8975.

Burgess, B. K., & Lowe, D. J. (2002). Mechanism of molybdenum nitrogenase. *Chemical Reviews, 96*, 2983–3012.

Callejas, J. F., McEnaney, J. M., Read, C. G., Crompton, J. C., Biacchi, A. J., Popczun, E. J., Gordon, T. R., Lewis, N. S., & Schaak, R. E. (2014). Electrocatalytic and photocatalytic hydrogen production from acidic and neutral-pH aqueous solutions using Iron phosphide nanoparticles. *ACS Nano, 8*, 11101–11107.

Callejas, J. F., Read, C. G., Popczun, E. J., McEnaney, J. M., & Schaak, R. E. (2015). Nanostructured Co2P electrocatalyst for the hydrogen evolution reaction and direct comparison with morphologically equivalent CoP. *Chemistry of Materials, 27*, 3769–3774.

Chen, S., Duan, J., Jaroniec, M., & Qiao, S.-Z. (2014). Nitrogen and oxygen dual-doped carbon hydrogel film as a substrate-free electrode for highly efficient oxygen evolution reaction. *Advanced Materials, 26*, 2925–2930.

Chen, S., Duan, J., Jaroniec, M., & Qiao, S. Z. (2013a). Three-dimensional N-doped graphene hydrogel/NiCo double hydroxide electrocatalysts for highly efficient oxygen evolution. *Angewandte Chemie International Edition, 52*, 13567–13570.

Chen, S., Duan, J., Ran, J., Jaroniec, M., & Qiao, S. Z. (2013b). N-doped graphene film-confined nickel nanoparticles as a highly efficient three-dimensional oxygen evolution electrocatalyst. *Energy & Environmental Science, 6*, 3693–3699.

Chen, S., & Qiao, S. Z. (2013). Hierarchically porous nitrogen-doped graphene-NiCo2O4 hybrid paper as an advanced electrocatalytic water-splitting material. *ACS Nano, 7*, 10190–10196.

Chen, X., Yu, K., Shen, Y., Feng, Y., & Zhu, Z. (2017). Synergistic effect of MoS2 nanosheets and VS2 for the hydrogen evolution reaction with enhanced humidity-sensing performance. *ACS Applied Materials & Interfaces, 9*, 42139–42148.

Chen, Z., Cummins, D., Reinecke, B. N., Clark, E., Sunkara, M. K., & Jaramillo, T. F. (2011). Core–shell MoO3-MoS2 nanowires for hydrogen evolution: A functional design for electrocatalytic materials. *Nano Letters, 11*, 4168–4175.

Chen, Z., Kronawitter, C. X., & Koel, B. E. (2015). Facet-dependent activity and stability of Co3O4 nanocrystals towards the oxygen evolution reaction. *Physical Chemistry Chemical Physics, 17*, 29387–29393.

Choi, C. L., Feng, J., Li, Y., Wu, J., Zak, A., Tenne, R., & Dai, H. (2013). WS2 nanoflakes from nanotubes for electrocatalysis. *Nano Research, 6*, 921–928.

Chung, D. Y., Han, J. W., Lim, D.-H., Jo, J.-H., Yoo, S. J., Lee, H., & Sung, Y.-E. (2015). Structure dependent active sites of NixSy as electrocatalysts for hydrogen evolution reaction. *Nanoscale, 7*, 5157–5163.

Chung, D. Y., Jun, S. W., Yoon, G., Kim, H., Yoo, J. M., Lee, K.-S., Kim, T., Shin, H., Sinha, A. K., Kwon, S. G., Kang, K., Hyeon, T., & Sung, Y.-E. (2017). Large-scale synthesis of carbon-shell-coated FeP nanoparticles for robust hydrogen evolution reaction electrocatalyst. *Journal of the American Chemical Society, 139*, 6669–6674.

Chung, D. Y., Park, S.-K., Chung, Y.-H., Yu, S.-H., Lim, D.-H., Jung, N., Ham, H. C., Park, H.-Y., Piao, Y., Yoo, S. J., & Sung, Y.-E. (2014). Edge-exposed MoS2 nano-assembled structures as efficient electrocatalysts for hydrogen evolution reaction. *Nanoscale, 6*, 2131–2136.

Das, J. K., Samantara, A. K., Nayak, A. K., Pradhan, D., & Behera, J. N. (2018). VS2: An efficient catalyst for an electrochemical hydrogen evolution reaction in an acidic medium. *Dalton Transactions, 47*, 13792–13799.

Di Giovanni, C., Wang, W.-A., Nowak, S., Grenèche, J.-M., Lecoq, H., Mouton, L., Giraud, M., & Tard, C. (2014). Bioinspired iron sulfide nanoparticles for cheap and long-lived electrocatalytic molecular hydrogen evolution in neutral water. *ACS Catalysis, 4*, 681–687.

Diaz-Morales, O., Raaijman, S., Kortlever, R., Kooyman, P. J., Wezendonk, T., Gascon, J., Fu, W. T., & Koper, M. T. M. (2016). Iridium-based double perovskites for efficient water oxidation in acid media. *Nature Communications, 7*, 12363.

Duan, J., Chen, S., Vasileff, A., & Qiao, S. Z. (2016). Anion and cation modulation in metal compounds for bifunctional overall water splitting. *ACS Nano, 10*, 8738–8745.

Dutta, A., Mutyala, S., Samantara, A. K., Bera, S., Jena, B. K., & Pradhan, N. (2018). Synergistic effect of inactive iron oxide core on active nickel phosphide shell for significant enhancement in oxygen evolution reaction activity. *ACS Energy Letters, 3*, 141–148.

Dutta, A., Samantara, A. K., Dutta, S. K., Jena, B. K., & Pradhan, N. (2016). Surface-oxidized dicobalt phosphide nanoneedles as a nonprecious, durable, and efficient OER catalyst. *ACS Energy Letters, 1*, 169–174.

Eady, R. R. (1996). Structure-function relationships of alternative nitrogenases. *Chemical Reviews, 96*, 3013–3030.

Eckenhoff, W. T., McNamara, W. R., Du, P., & Eisenberg, R. (2013). Cobalt complexes as artificial hydrogenases for the reductive side of water splitting. *Biochimica et Biophysica Acta (BBA) - Bioenergetics, 1827*, 958–973.

Ensafi, A. A., Jafari-Asl, M., Nabiyan, A., & Rezaei, B. (2016). Ni3S2/ball-milled silicon flour as a bi-functional electrocatalyst for hydrogen and oxygen evolution reactions. *Energy, 116*, 392–401.

Esposito, D. V., & Chen, J. G. (2011). Monolayer platinum supported on tungsten carbides as low-cost electrocatalysts: Opportunities and limitations. *Energy & Environmental Science, 4*, 3900–3912.

Esswein, A. J., McMurdo, M. J., Ross, P. N., Bell, A. T., & Tilley, T. D. (2009). Size-dependent activity of Co3O4 nanoparticle anodes for alkaline water electrolysis. *Journal of Physical Chemistry C, 113*, 15068–15072.

Faber, M. S., Dziedzic, R., Lukowski, M. A., Kaiser, N. S., Ding, Q., & Jin, S. (2014a). High-performance electrocatalysis using metallic cobalt pyrite (CoS2) micro- and nanostructures. *Journal of the American Chemical Society, 136*, 10053–10061.

Faber, M. S., Lukowski, M. A., Ding, Q., Kaiser, N. S., & Jin, S. (2014b). Earth-abundant metal pyrites (FeS2, CoS2, NiS2, and their alloys) for highly efficient hydrogen evolution and polysulfide reduction Electrocatalysis. *Journal of Physical Chemistry C, 118*, 21347–21356.

Feng, L.-L., Yu, G., Wu, Y., Li, G.-D., Li, H., Sun, Y., Asefa, T., Chen, W., & Zou, X. (2015). High-index faceted Ni3S2 nanosheet arrays as highly active and ultrastable electrocatalysts for water splitting. *Journal of the American Chemical Society, 137*, 14023–14026.

Feng, Y., He, T., & Alonso-Vante, N. (2008). In situ free-surfactant synthesis and ORR- electrochemistry of carbon-supported Co3S4 and CoSe2 nanoparticles. *Chemistry of Materials, 20*, 26–28.

Gao, M., Sheng, W., Zhuang, Z., Fang, Q., Gu, S., Jiang, J., & Yan, Y. (2014). Efficient water oxidation using nanostructured α-nickel-hydroxide as an electrocatalyst. *Journal of the American Chemical Society, 136*, 7077–7084.

Gilje, S., Kaner, R. B., Wallace, G. G., Li, D. A. N., & Mu, M. B. (2008). Processable aqueous dispersions of graphene nanosheets. *Nature Nanotechnology, 3*, 101–105.

Gong, M., Li, Y., Wang, H., Liang, Y., Wu, J. Z., Zhou, J., Wang, J., Regier, T., Wei, F., & Dai, H. (2013). An advanced Ni–Fe layered double hydroxide electrocatalyst for water oxidation. *Journal of the American Chemical Society, 135*, 8452–8455.

Gong, M., Zhou, W., Tsai, M.-C., Zhou, J., Guan, M., Lin, M.-C., Zhang, B., Hu, Y., Wang, D.-Y., Yang, J., Pennycook, S. J., Hwang, B.-J., & Dai, H. (2014). Nanoscale nickel oxide/nickel heterostructures for active hydrogen evolution electrocatalysis. *Nature Communications, 5*, 4695.

Gopalakrishnan, D., Damien, D., & Shaijumon, M. M. (2014). MoS2 quantum dot-interspersed exfoliated MoS2 Nanosheets. *ACS Nano, 8*, 5297–5303.

Greeley, J., Jaramillo, T. F., Bonde, J., Chorkendorff, I., & Nørskov, J. K. (2006). Computational high-throughput screening of electrocatalytic materials for hydrogen evolution. *Nature Materials, 5*, 909–913.

Guo, C. X., Chen, S., & Lu, X. (2014). Ethylenediamine-mediated synthesis of Mn3O4 nano-octahedrons and their performance as electrocatalysts for the oxygen evolution reaction. *Nanoscale, 6*, 10896–10901.

Guo, Y., Tong, Y., Chen, P., Xu, K., Zhao, J., Lin, Y., Chu, W., Peng, Z., Wu, C., & Xie, Y. (2015). Engineering the electronic state of a Perovskite electrocatalyst for synergistically enhanced oxygen evolution reaction. *Advanced Materials, 27*, 5989–5994.

Hallenbeck, P. C., & Benemann, J. R. (2002). Biological hydrogen production; fundamentals and limiting processes. *International Journal of Hydrogen Energy, 27*, 1185–1193.

Hinnemann, B., Moses, P. G., Bonde, J., Jørgensen, K. P., Nielsen, J. H., Horch, S., Chorkendorff, I., & Nørskov, J. K. (2005). Biomimetic hydrogen evolution: MoS2 nanoparticles as catalyst for hydrogen evolution. *Journal of the American Chemical Society, 127*, 5308–5309.

Hirai, S., Yagi, S., Seno, A., Fujioka, M., Ohno, T., & Matsuda, T. (2016). Enhancement of the oxygen evolution reaction in Mn3+−based electrocatalysts: Correlation between Jahn–Teller distortion and catalytic activity. *RSC Advances, 6*, 2019–2023.

Hüppauff, M. (1993). Valency and structure of iridium in anodic iridium oxide films. *Journal of the Electrochemical Society, 140*, 598.

Hutchings, R., Müller, K., Kötz, R., & Stucki, S. (1984). A structural investigation of stabilized oxygen evolution catalysts. *Journal of Materials Science, 19*, 3987–3994.

Huynh, M., Shi, C., Billinge, S. J. L., & Nocera, D. G. (2015). Nature of activated manganese oxide for oxygen evolution. *Journal of the American Chemical Society, 137*, 14887–14904.

Jaramillo, T. F., Jørgensen, K. P., Bonde, J., Nielsen, J. H., Horch, S., & Chorkendorff, I. (2007). Identification of active edge sites for electrochemical H2 evolution from MoS2 nanocatalysts. *Science (80-.), 317*, 100–102.

Jiang, J., Wang, C., Zhang, J., Wang, W., Zhou, X., Pan, B., Tang, K., Zuo, J., & Yang, Q. (2015). Synthesis of FeP2/C nanohybrids and their performance for hydrogen evolution reaction. *Journal of Materials Chemistry A, 3*, 499–503.

Jiang, N., Bogoev, L., Popova, M., Gul, S., Yano, J., & Sun, Y. (2014). Electrodeposited nickel-sulfide films as competent hydrogen evolution catalysts in neutral water. *Journal of Materials Chemistry A, 2*, 19407–19414.

Jiang, P., Liu, Q., Ge, C., Cui, W., Pu, Z., Asiri, A. M., & Sun, X. (2014a). CoP nanostructures with different morphologies: Synthesis, characterization and a study of their electrocatalytic performance toward the hydrogen evolution reaction. *Journal of Materials Chemistry A, 2*, 14634–14640.

Jiang, P., Liu, Q., Liang, Y., Tian, J., Asiri, A. M., & Sun, X. (2014b). A cost-effective 3D hydrogen evolution cathode with high catalytic activity: FeP nanowire array as the active phase. *Angewandte Chemie, 126*, 13069–13073.

Jiang, P., Liu, Q., & Sun, X. (2014c). NiP2 nanosheet arrays supported on carbon cloth: An efficient 3D hydrogen evolution cathode in both acidic and alkaline solutions. *Nanoscale, 6*, 13440–13445.

Jing, S., Lu, J., Yu, G., Yin, S., Luo, L., Zhang, Z., Ma, Y., Chen, W., & Shen, P. K. (2018). Carbon-encapsulated WOx hybrids as efficient catalysts for hydrogen evolution. *Advanced Materials, 30*, 1705979.

Kelly, T. G., & Chen, J. G. (2012). Metal overlayer on metal carbide substrate: Unique bimetallic properties for catalysis and electrocatalysis. *Chemical Society Reviews, 41*, 8021–8034.

Kibler, L. A., El-Aziz, A. M., Hoyer, R., & Kolb, D. M. (2005). Tuning reaction rates by lateral strain in a palladium monolayer. *Angewandte Chemie International Edition, 44*, 2080–2084.

Kibsgaard, J., Chen, Z., Reinecke, B. N., & Jaramillo, T. F. (2012). Engineering the surface structure of MoS2 to preferentially expose active edge sites for electrocatalysis. *Nature Materials, 11*, 963.

Kibsgaard, J., Tsai, C., Chan, K., Benck, J. D., Nørskov, J. K., Abild-Pedersen, F., & Jaramillo, T. F. (2015). Designing an improved transition metal phosphide catalyst for hydrogen evolution using experimental and theoretical trends. *Energy & Environmental Science, 8*, 3022–3029.

Kim, J., Kim, J. S., Baik, H., Kang, K., & Lee, K. (2016). Porous β-MnO2 nanoplates derived from MnCO3 nanoplates as highly efficient electrocatalysts toward oxygen evolution reaction. *RSC Advances, 6*, 26535–26539.

Kim, N. -I., Sa, Y. J., Cho, S. -H., So, I., Kwon, K., Joo, S. H., & Park, J. -Y. (2016). Enhancing activity and stability of cobalt oxide electrocatalysts for the oxygen evolution reaction via transition metal doping. *Journal of the Electrochemical Society, 163*, F3020–F3028.

Kitchin, J. R., Nørskov, J. K., Barteau, M. A., & Chen, J. G. (2004). Role of strain and ligand effects in the modification of the electronic and chemical properties of bimetallic surfaces. *Physical Review Letters, 93*, 156801.

Kong, D., Cha, J. J., Wang, H., Lee, H. R., & Cui, Y. (2013a). First-row transition metal dichalcogenide catalysts for hydrogen evolution reaction. *Energy & Environmental Science, 6*, 3553–3558.

Kong, D., Wang, H., Cha, J. J., Pasta, M., Koski, K. J., Yao, J., & Cui, Y. (2013b). Synthesis of MoS2 and MoSe2 films with vertically aligned layers. *Nano Letters, 13*, 1341–1347.

Kong, D., Wang, H., Lu, Z., & Cui, Y. (2014). CoSe2 nanoparticles grown on carbon fiber paper: An efficient and stable electrocatalyst for hydrogen evolution reaction. *Journal of the American Chemical Society, 136*, 4897–4900.

Kötz, R., & Stucki, S. (1985). Oxygen evolution and corrosion on ruthenium-iridium alloys. *Journal of the Electrochemical Society, 132*, 103–107.

Kreysa, G., Ota, K.-I., & Savinell, R. F. (2014). *Encyclopedia of applied electrochemistry*. New York: Springer.

Kuo, C.-H., Li, W., Pahalagedara, L., El-Sawy, A. M., Kriz, D., Genz, N., Guild, C., Ressler, T., Suib, S. L., & He, J. (2015). Understanding the role of gold nanoparticles in enhancing the catalytic activity of manganese oxides in water oxidation reactions. *Angewandte Chemie International Edition, 54*, 2345–2350.

Laursen, A. B., Patraju, K. R., Whitaker, M. J., Retuerto, M., Sarkar, T., Yao, N., Ramanujachary, K. V., Greenblatt, M., & Dismukes, G. C. (2015). Nanocrystalline Ni5P4: A hydrogen evolution electrocatalyst of exceptional efficiency in both alkaline and acidic media. *Energy & Environmental Science, 8*, 1027–1034.

Laursen, A. B., Wexler, R. B., Whitaker, M. J., Izett, E. J., Calvinho, K. U. D., Hwang, S., Rucker, R., Wang, H., Li, J., Garfunkel, E., Greenblatt, M., Rappe, A. M., & Dismukes, G. C. (2018). Climbing the volcano of electrocatalytic activity while avoiding catalyst corrosion: Ni3P, a hydrogen evolution electrocatalyst stable in both acid and alkali. *ACS Catalysis, 8*, 4408–4419.

Lee, D. U., Choi, J.-Y., Feng, K., Park, H. W., & Chen, Z. (2014). Advanced extremely durable 3D bifunctional air electrodes for rechargeable zinc-air batteries. *Advanced Energy Materials, 4*, 1301389.

Lee, Y., Suntivich, J., May, K. J., Perry, E. E., & Shao-Horn, Y. (2012). Synthesis and activities of rutile IrO2 and RuO2 nanoparticles for oxygen evolution in acid and alkaline solutions. *Journal of Physical Chemistry Letters, 3*, 399–404.

Leng, X., Zeng, Q., Wu, K.-H., Gentle, I. R., & Wang, D.-W. (2015). Reduction-induced surface amorphization enhances the oxygen evolution activity in Co3O4. *RSC Advances, 5*, 27823–27828.

Li, D., Baydoun, H., Verani, C. N., & Brock, S. L. (2016). Efficient water oxidation using CoMnP nanoparticles. *Journal of the American Chemical Society, 138*, 4006–4009.

Li, M., Xiong, Y., Liu, X., Bo, X., Zhang, Y., Han, C., & Guo, L. (2015). Facile synthesis of electrospun MFe2O4 (M = Co, Ni, Cu, Mn) spinel nanofibers with excellent electrocatalytic properties for oxygen evolution and hydrogen peroxide reduction. *Nanoscale, 7*, 8920–8930.

Li, S., Wang, Y., Peng, S., Zhang, L., Al-Enizi, A. M., Zhang, H., Sun, X., & Zheng, G. (2016). Co–Ni-based nanotubes/Nanosheets as efficient water splitting Electrocatalysts. *Advanced Energy Materials, 6*, 1501661.

Li, Y., Zhang, H., Jiang, M., Kuang, Y., Sun, X., & Duan, X. (2016). Ternary NiCoP nanosheet arrays: An excellent bifunctional catalyst for alkaline overall water splitting. *Nano Research, 9*, 2251–2259.

Li, Y. H., Liu, P. F., Pan, L. F., Wang, H. F., Yang, Z. Z., Zheng, L. R., Hu, P., Zhao, H. J., Gu, L., & Yang, H. G. (2015). Local atomic structure modulations activate metal oxide as electrocatalyst for hydrogen evolution in acidic water. *Nature Communications, 6*, 8064.

Liang, H., Shi, H., Zhang, D., Ming, F., Wang, R., Zhuo, J., & Wang, Z. (2016). Solution growth of vertical VS2 nanoplate arrays for electrocatalytic hydrogen evolution. *Chemistry of Materials, 28*, 5587–5591.

Liang, Y., Li, Y., Wang, H., Zhou, J., Wang, J., Regier, T., & Dai, H. (2011). Co3O4 nanocrystals on graphene as a synergistic catalyst for oxygen reduction reaction. *Nature Materials, 10*, 780.

Liao, J.-Y., Higgins, D., Lui, G., Chabot, V., Xiao, X., & Chen, Z. (2013). Multifunctional TiO2–C/MnO2 Core–double-shell nanowire arrays as high-performance 3D electrodes for lithium ion batteries. *Nano Letters, 13*, 5467–5473.

Lin, C.-H., Chen, C.-L., & Wang, J.-H. (2011). Mechanistic studies of water–gas-shift reaction on transition metals. *Journal of Physical Chemistry C, 115*, 18582–18588.

Liu, P., & Rodriguez, J. A. (2005). Catalysts for hydrogen evolution from the [NiFe] hydrogenase to the Ni2P(001) surface: The importance of ensemble effect. *Journal of the American Chemical Society, 127*, 14871–14878.

Liu, Q., Tian, J., Cui, W., Jiang, P., Cheng, N., Asiri, A. M., & Sun, X. (2014). Carbon nanotubes decorated with CoP nanocrystals: A highly active non-noble-metal nanohybrid electrocatalyst for hydrogen evolution. *Angewandte Chemie International Edition, 53*, 6710–6714.

Liu, T., Liang, Y., Liu, Q., Sun, X., He, Y., & Asiri, A. M. (2015). Electrodeposition of cobalt-sulfide nanosheets film as an efficient electrocatalyst for oxygen evolution reaction. *Electrochemistry Communications, 60*, 92–96.

Liu, W., Hu, E., Jiang, H., Xiang, Y., Weng, Z., Li, M., Fan, Q., Yu, X., Altman, E. I., & Wang, H. (2016). A highly active and stable hydrogen evolution catalyst based on pyrite-structured cobalt phosphosulfide. *Nature Communications, 7*, 10771.

Liu, X., Cui, S., Qian, M., Sun, Z., & Du, P. (2016). In situ generated highly active copper oxide catalysts for the oxygen evolution reaction at low overpotential in alkaline solutions. *Chemical Communications, 52*, 5546–5549.

Liu, Y., Cheng, H., Lyu, M., Fan, S., Liu, Q., Zhang, W., Zhi, Y., Wang, C., Xiao, C., Wei, S., Ye, B., & Xie, Y. (2014). Low overpotential in vacancy-rich ultrathin CoSe2 nanosheets for water oxidation. *Journal of the American Chemical Society, 136*, 15670–15675.

Lu, B., Cao, D., Wang, P., Wang, G., & Gao, Y. (2011). Oxygen evolution reaction on Ni-substituted Co3O4 nanowire array electrodes. *International Journal of Hydrogen Energy, 36*, 72–78.

Lu, Z., Zhang, H., Zhu, W., Yu, X., Kuang, Y., Chang, Z., Lei, X., & Sun, X. (2013). In situ fabrication of porous MoS2 thin-films as high-performance catalysts for electrochemical hydrogen evolution. *Chemical Communications, 49*, 7516–7518.

Lu, Z., Zhu, W., Yu, X., Zhang, H., Li, Y., Sun, X., Wang, X., Wang, H., Wang, J., Luo, J., Lei, X., & Jiang, L. (2014b). Ultrahigh hydrogen evolution performance of under-water "Superaerophobic" MoS2 nanostructured electrodes. *Advanced Materials, 26*, 2683–2687.

Lukowski, M. A., Daniel, A. S., Meng, F., Forticaux, A., Li, L., & Jin, S. (2013). Enhanced hydrogen evolution catalysis from chemically exfoliated metallic MoS2 nanosheets. *Journal of the American Chemical Society, 135*, 10274–10277.

Lv, X.-J., She, G.-W., Zhou, S.-X., & Li, Y.-M. (2013). Highly efficient electrocatalytic hydrogen production by nickel promoted molybdenum sulfide microspheres catalysts. *RSC Advances, 3*, 21231–21236.

Ma, T. Y., Dai, S., & Qiao, S. Z. (2016). Self-supported electrocatalysts for advanced energy conversion processes. *Materials Today, 19*, 265–273.

Ma, T. Y., Ran, J., Dai, S., Jaroniec, M., & Qiao, S. Z. (2015). Phosphorus-doped graphitic carbon nitrides grown in situ on carbon-fiber paper: Flexible and reversible oxygen electrodes. *Angewandte Chemie International Edition, 54*, 4646–4650.

Mamaca, N., Mayousse, E., Arrii-Clacens, S., Napporn, T. W., Servat, K., Guillet, N., & Kokoh, K. B. (2012). Electrochemical activity of ruthenium and iridium based catalysts for oxygen evolution reaction. *Applied Catalysis B: Environmental, 111–112*, 376–380.

Man, I. C., Su, H.-Y., Calle-Vallejo, F., Hansen, H. A., Martínez, J. I., Inoglu, N. G., Kitchin, J., Jaramillo, T. F., Nørskov, J. K., & Rossmeisl, J. (2011). Universality in oxygen evolution electrocatalysis on oxide surfaces. *ChemCatChem, 3*, 1159–1165.

McEnaney, J. M., Chance Crompton, J., Callejas, J. F., Popczun, E. J., Read, C. G., Lewis, N. S., & Schaak, R. E. (2014). Electrocatalytic hydrogen evolution using amorphous tungsten phosphide nanoparticles. *Chemical Communications, 50*, 11026–11028.

McPherson, I. J., & Vincent, K. A. (2014). Electrocatalysis by hydrogenases: Lessons for building bio-inspired devices. *Journal of the Brazilian Chemical Society, 25*, 427–441.

Mendoza-Garcia, A., Zhu, H., Yu, Y., Li, Q., Zhou, L., Su, D., Kramer, M. J., & Sun, S. (2015). Controlled anisotropic growth of Co-Fe-P from Co-Fe-O nanoparticles. *Angewandte Chemie International Edition, 54*, 9642–9645.

Merki, D., Vrubel, H., Rovelli, L., Fierro, S., & Hu, X. (2012). Fe, co, and Ni ions promote the catalytic activity of amorphous molybdenum sulfide films for hydrogen evolution. *Chemical Science, 3*, 2515–2525.

Mom, R. V., Cheng, J., Koper, M. T. M., & Sprik, M. (2014). Modeling the oxygen evolution reaction on metal oxides: The Infuence of unrestricted DFT calculations. *Journal of Physical Chemistry C, 118*, 4095–4102.

Muthuswamy, E., Kharel, P. R., Lawes, G., & Brock, S. L. (2009). Control of phase in phosphide nanoparticles produced by metal nanoparticle transformation: Fe2P and FeP. *ACS Nano, 3*, 2383–2393.

Nahor, G. S., Hapiot, P., Neta, P., & Harriman, A. (1991). Changes in the redox state of iridium oxide clusters and their relation to catalytic water oxidation: Radiolytic and electrochemical studies. *The Journal of Physical Chemistry, 95*, 616–621.

Neyerlin, K. C., Bugosh, G., Forgie, R., Liu, Z., & Strasser, P. (2009). Combinatorial study of high-surface-area binary and ternary electrocatalysts for the oxygen evolution reaction. *Journal of the Electrochemical Society, 156*, B363–B369.

Ouyang, C., Wang, X., Wang, C., Zhang, X., Wu, J., Ma, Z., Dou, S., & Wang, S. (2015). Hierarchically porous Ni3S2 nanorod array foam as highly efficient electrocatalyst for hydrogen evolution reaction and oxygen evolution reaction. *Electrochimica Acta, 174*, 297–301.

Over, H. (2012). Surface chemistry of ruthenium dioxide in heterogeneous catalysis and electrocatalysis: From fundamental to applied research. *Chemical Reviews, 112*, 3356–3426.

Pan, Y., Liu, Y., Zhao, J., Yang, K., Liang, J., Liu, D., Hu, W., Liu, D., Liu, Y., & Liu, C. (2015). Monodispersed nickel phosphide nanocrystals with different phases: Synthesis, characterization and electrocatalytic properties for hydrogen evolution. *Journal of Materials Chemistry A, 3*, 1656–1665.

Pei, J., Mao, J., Liang, X., Chen, C., Peng, Q., Wang, D., & Li, Y. (2016). Ir–Cu nanoframes: One-pot synthesis and efficient electrocatalysts for oxygen evolution reaction. *Chemical Communications, 52*, 3793–3796.

Pi, M., Wu, T., Zhang, D., Chen, S., & Wang, S. (2016). Self-supported three-dimensional mesoporous semimetallic WP2 nanowire arrays on carbon cloth as a flexible cathode for efficient hydrogen evolution. *Nanoscale, 8*, 19779–19786.

Plaisance, C. P., Reuter, K., & van Santen, R. A. (2016). Quantum chemistry of the oxygen evolution reaction on cobalt(ii,iii) oxide – implications for designing the optimal catalyst. *Faraday Discussions, 188*, 199–226.

Popczun, E. J., McKone, J. R., Read, C. G., Biacchi, A. J., Wiltrout, A. M., Lewis, N. S., & Schaak, R. E. (2013). Nanostructured nickel phosphide as an electrocatalyst for the hydrogen evolution reaction. *Journal of the American Chemical Society, 135*, 9267–9270.

Popczun, E. J., Read, C. G., Roske, C. W., Lewis, N. S., & Schaak, R. E. (2014). Highly active electrocatalysis of the hydrogen evolution reaction by cobalt phosphide nanoparticles. *Angewandte Chemie International Edition, 53*, 5427–5430.

Pu, Z., Liu, Q., Asiri, A. M., Obaid, A. Y., & Sun, X. (2014a). One-step electrodeposition fabrication of graphene film-confined WS2 nanoparticles with enhanced electrochemical catalytic activity for hydrogen evolution. *Electrochimica Acta, 134*, 8–12.

Pu, Z., Liu, Q., Asiri, A. M., & Sun, X. (2014b). Tungsten phosphide nanorod arrays directly grown on carbon cloth: A highly efficient and stable hydrogen evolution cathode at all pH values. *ACS Applied Materials & Interfaces, 6*, 21874–21879.

Pu, Z., Ya, X., Amiinu, I. S., Tu, Z., Liu, X., Li, W., & Mu, S. (2016). Ultrasmall tungsten phosphide nanoparticles embedded in nitrogen-doped carbon as a highly active and stable hydrogen-evolution electrocatalyst. *Journal of Materials Chemistry A, 4*, 15327–15332.

Qiu, Y., Xin, L., & Li, W. (2014). Electrocatalytic oxygen evolution over supported small amorphous Ni–Fe nanoparticles in alkaline electrolyte. *Langmuir, 30*, 7893–7901.

Rao, Y., Zhang, L.-M., Shang, X., Dong, B., Liu, Y.-R., Lu, S.-S., Chi, J.-Q., Chai, Y.-M., & Liu, C.-G. (2017). Vanadium sulfides interwoven nanoflowers based on in-situ sulfurization of vanadium oxides octahedron on nickel foam for efficient hydrogen evolution. *Applied Surface Science, 423*, 1090–1096.

Ratha, S., Samantara, A. K., Singha, K. K., Gangan, A. S., Chakraborty, B., Jena, B. K., & Rout, C. S. (2017). Urea-assisted room temperature stabilized metastable β-NiMoO4: Experimental and theoretical insights into its unique Bifunctional activity toward oxygen evolution and Supercapacitor. *ACS Applied Materials & Interfaces, 9*, 9640–9653.

Ryu, J., Jung, N., Jang, J. H., Kim, H.-J., & Yoo, S. J. (2015). In situ transformation of hydrogen-evolving CoP nanoparticles: Toward efficient oxygen evolution catalysts bearing dispersed morphologies with Co-oxo/hydroxo molecular units. *ACS Catalysis, 5*, 4066–4074.

Samantara, A. K., Kamila, S., Ghosh, A., & Jena, B. K. (2018). Highly ordered 1D NiCo2O4 nanorods on graphene: An efficient dual-functional hybrid materials for electrochemical energy conversion and storage applications. *Electrochimica Acta, 263*, 147–157.

Sanchez Casalongue, H. G., Ng, M. L., Kaya, S., Friebel, D., Ogasawara, H., & Nilsson, A. (2014). In situ observation of surface species on iridium oxide nanoparticles during the oxygen evolution reaction. *Angewandte Chemie International Edition, 53*, 7169–7172.

Sardar, K., Petrucco, E., Hiley, C. I., Sharman, J. D. B., Wells, P. P., Russell, A. E., Kashtiban, R. J., Sloan, J., & Walton, R. I. (2014). Water-splitting electrocatalysis in acid conditions using Ruthenate-Iridate Pyrochlores. *Angewandte Chemie International Edition, 53*, 10960–10964.

Schipper, D. E., Zhao, Z., Thirumalai, H., Leitner, A. P., Donaldson, S. L., Kumar, A., Qin, F., Wang, Z., Grabow, L. C., Bao, J., & Whitmire, K. H. (2018). Effects of catalyst phase on the hydrogen evolution reaction of water splitting: Preparation of phase-pure films of FeP, Fe2P, and Fe3P and their relative catalytic activities. *Chemistry of Materials, 30*, 3588–3598.

Seo, B., Baek, D. S., Sa, Y. J., & Joo, S. H. (2016). Shape effects of nickel phosphide nanocrystals on hydrogen evolution reaction. *CrystEngComm, 18*, 6083–6089.

Shen, M., Ruan, C., Chen, Y., Jiang, C., Ai, K., & Lu, L. (2015). Covalent entrapment of cobalt–iron sulfides in N-doped Mesoporous carbon: Extraordinary bifunctional electrocatalysts for oxygen reduction and evolution reactions. *ACS Applied Materials & Interfaces, 7*, 1207–1218.

Smith, R. D. L., Prévot, M. S., Fagan, R. D., Trudel, S., & Berlinguette, C. P. (2013). Water oxidation catalysis: Electrocatalytic response to metal stoichiometry in amorphous metal oxide films containing Iron, cobalt, and nickel. *Journal of the American Chemical Society, 135*, 11580–11586.

Song, F., & Hu, X. (2014). Exfoliation of layered double hydroxides for enhanced oxygen evolution catalysis. *Nature Communications, 5*, 4477.

Song, F., Schenk, K., & Hu, X. (2016). A nanoporous oxygen evolution catalyst synthesized by selective electrochemical etching of perovskite hydroxide CoSn(OH)6 nanocubes. *Energy & Environmental Science, 9*, 473–477.

Subbaraman, R., Tripkovic, D., Chang, K.-C., Strmcnik, D., Paulikas, A. P., Hirunsit, P., Chan, M., Greeley, J., Stamenkovic, V., & Markovic, N. M. (2012). Trends in activity for the water electrolyser reactions on 3d M(Ni,Co,Fe,Mn) hydr(oxy)oxide catalysts. *Nature Materials, 11*, 550.

Sun, X., Dai, J., Guo, Y., Wu, C., Hu, F., Zhao, J., Zeng, X., & Xie, Y. (2014). Semimetallic molybdenum disulfide ultrathin nanosheets as an efficient electrocatalyst for hydrogen evolution. *Nanoscale, 6*, 8359–8367.

Suntivich, J., May, K. J., Gasteiger, H. A., Goodenough, J. B., & Shao-Horn, Y. (2011). A Perovskite oxide optimized for oxygen evolution catalysis from molecular orbital principles. *Science (80-.), 334*, 1383 LP–1385.

Tang, C., Gan, L., Zhang, R., Lu, W., Jiang, X., Asiri, A. M., Sun, X., Wang, J., & Chen, L. (2016). Ternary FexCo1–xP nanowire array as a robust hydrogen evolution reaction electrocatalyst with Pt-like activity: Experimental and theoretical insight. *Nano Letters, 16*, 6617–6621.

Tang, R., Nie, Y., Kawasaki, J. K., Kuo, D. -Y., Petretto, G., Hautier, G., Rignanese, G. -M., Shen, K. M., Schlom, D. G., & Suntivich, J. (2016). Oxygen evolution reaction electrocatalysis on SrIrO3 grown using molecular beam epitaxy. *Journal of Materials Chemistry A, 4*, 6831–6836.

Tian, J., Liu, Q., Asiri, A. M., & Sun, X. (2014a). Self-supported nanoporous cobalt phosphide nanowire arrays: An efficient 3D hydrogen-evolving cathode over the wide range of pH 0–14. *Journal of the American Chemical Society, 136*, 7587–7590.

Tian, J., Liu, Q., Cheng, N., Asiri, A. M., & Sun, X. (2014b). Self-supported Cu3P nanowire arrays as an integrated high-performance three-dimensional cathode for generating hydrogen from water. *Angewandte Chemie, 126*, 9731–9735.

Tran, P. D., Chiam, S. Y., Boix, P. P., Ren, Y., Pramana, S. S., Fize, J., Artero, V., & Barber, J. (2013). Novel cobalt/nickel–tungsten-sulfide catalysts for electrocatalytic hydrogen generation from water. *Energy & Environmental Science, 6*, 2452–2459.

Trasatti, S. (1984). Electrocatalysis in the anodic evolution of oxygen and chlorine. *Electrochimica Acta, 29*, 1503–1512.

Trotochaud, L., Ranney, J. K., Williams, K. N., & Boettcher, S. W. (2012). Solution-cast metal oxide thin film electrocatalysts for oxygen evolution. *Journal of the American Chemical Society, 134*, 17253–17261.

Vasić Anićijević, D. D., Nikolić, V. M., Marčeta-Kaninski, M. P., & Pašti, I. A. (2013). Is platinum necessary for efficient hydrogen evolution? – DFT study of metal monolayers on tungsten carbide. *International Journal of Hydrogen Energy, 38*, 16071–16079.

Voiry, D., Salehi, M., Silva, R., Fujita, T., Chen, M., Asefa, T., Shenoy, V. B., Eda, G., & Chhowalla, M. (2013a). Conducting MoS2 nanosheets as catalysts for hydrogen evolution reaction. *Nano Letters, 13*, 6222–6227.

Voiry, D., Yamaguchi, H., Li, J., Silva, R., Alves, D. C. B., Fujita, T., Chen, M., Asefa, T., Shenoy, V. B., Eda, G., & Chhowalla, M. (2013b). Enhanced catalytic activity in strained chemically exfoliated WS2 nanosheets for hydrogen evolution. *Nature Materials, 12*, 850.

Wang, D., Pan, Z., Wu, Z., Wang, Z., & Liu, Z. (2014). Hydrothermal synthesis of MoS2 nanoflowers as highly efficient hydrogen evolution reaction catalysts. *Journal of Power Sources, 264*, 229–234.

Wang, D., Wang, Z., Wang, C., Zhou, P., Wu, Z., & Liu, Z. (2013). Distorted MoS2 nanostructures: An efficient catalyst for the electrochemical hydrogen evolution reaction. *Electrochemistry Communications, 34*, 219–222.

Wang, H., Kong, D., Johanes, P., Cha, J. J., Zheng, G., Yan, K., Liu, N., & Cui, Y. (2013). MoSe2 and WSe2 nanofilms with vertically aligned molecular layers on curved and rough surfaces. *Nano Letters, 13*, 3426–3433.

Wang, H., Lee, H. -W., Deng, Y., Lu, Z., Hsu, P. -C., Liu, Y., Lin, D., & Cui, Y. (2015). Bifunctional non-noble metal oxide nanoparticle electrocatalysts through lithium-induced conversion for overall water splitting. *Nature Communications, 6*, 7261.

Wang, R., Dong, X.-Y., Du, J., Zhao, J.-Y., & Zang, S.-Q. (2018). MOF-derived bifunctional Cu3P nanoparticles coated by a N,P-codoped carbon shell for hydrogen evolution and oxygen reduction. *Advanced Materials, 30*, 1703711.

Wang, X., Kolen'ko, Y. V., Bao, X.-Q., Kovnir, K., & Liu, L. (2015). One-step synthesis of self-supported nickel phosphide nanosheet array cathodes for efficient electrocatalytic hydrogen generation. *Angewandte Chemie, 127*, 8306–8310.

Wang, Y., Zhou, T., Jiang, K., Da, P., Peng, Z., Tang, J., Kong, B., Cai, W. -B., Yang, Z., & Zheng, G. (2014). Electrocatalysis: Reduced Mesoporous Co3O4 nanowires as efficient water oxidation electrocatalysts and supercapacitor electrodes (Adv. Energy Mater. 16/2014). *Advanced Energy Materials, 4*: 1400696.

Wu, T., Pi, M., Wang, X., Guo, W., Zhang, D., & Chen, S. (2017a). Developing bifunctional electrocatalyst for overall water splitting using three-dimensional porous CoP3 nanospheres integrated on carbon cloth. *Journal of Alloys and Compounds, 729*, 203–209.

Wu, T., Pi, M., Wang, X., Zhang, D., & Chen, S. (2017b). Three-dimensional metal–organic framework derived porous CoP3 concave polyhedrons as superior bifunctional electrocatalysts for the evolution of hydrogen and oxygen. *Physical Chemistry Chemical Physics, 19*, 2104–2110.

Wu, X., & Scott, K. (2013). A Li-doped Co3O4 oxygen evolution catalyst for non-precious metal alkaline anion exchange membrane water electrolysers. *International Journal of Hydrogen Energy, 38*, 3123–3129.

Wu, Z., Fang, B., Wang, Z., Wang, C., Liu, Z., Liu, F., Wang, W., Alfantazi, A., Wang, D., & Wilkinson, D. P. (2013). MoS2 nanosheets: A designed structure with high active site density for the hydrogen evolution reaction. *ACS Catalysis, 3*, 2101–2107.

Xia, X., Figueroa-Cosme, L., Tao, J., Peng, H.-C., Niu, G., Zhu, Y., & Xia, Y. (2014). Facile synthesis of iridium nanocrystals with well-controlled facets using seed-mediated growth. *Journal of the American Chemical Society, 136*, 10878–10881.

Xiao, P., Sk, M. A., Thia, L., Ge, X., Lim, R. J., Wang, J.-Y., Lim, K. H., & Wang, X. (2014). Molybdenum phosphide as an efficient electrocatalyst for the hydrogen evolution reaction. *Energy & Environmental Science, 7*, 2624–2629.

Xie, J., Zhang, J., Li, S., Grote, F., Zhang, X., Zhang, H., Wang, R., Lei, Y., Pan, B., & Xie, Y. (2013). Controllable disorder engineering in oxygen-incorporated MoS2 ultrathin Nanosheets for efficient hydrogen evolution. *Journal of the American Chemical Society, 135*, 17881–17888.

Xing, Z., Liu, Q., Asiri, A. M., & Sun, X. (2014). Closely interconnected network of molybdenum phosphide nanoparticles: A highly efficient Electrocatalyst for generating hydrogen from water. *Advanced Materials, 26*, 5702–5707.

Xing, Z., Liu, Q., Asiri, A. M., & Sun, X. (2015). High-efficiency electrochemical hydrogen evolution catalyzed by tungsten phosphide submicroparticles. *ACS Catalysis, 5*, 145–149.

Yan, X., Tian, L., He, M., & Chen, X. (2015). Three-dimensional crystalline/amorphous Co/Co3O4 core/shell nanosheets as efficient Electrocatalysts for the hydrogen evolution reaction. *Nano Letters, 15*, 6015–6021.

Yang, H., Zhang, Y., Hu, F., & Wang, Q. (2015). Urchin-like CoP nanocrystals as hydrogen evolution reaction and oxygen reduction reaction dual-electrocatalyst with superior stability. *Nano Letters, 15*, 7616–7620.

Yang, J., Voiry, D., Ahn, S. J., Kang, D., Kim, A. Y., Chhowalla, M., & Shin, H. S. (2013). Two-dimensional hybrid nanosheets of tungsten disulfide and reduced graphene oxide as catalysts for enhanced hydrogen evolution. *Angewandte Chemie International Edition, 52*, 13751–13754.

Yang, J., Zhang, F., Wang, X., He, D., Wu, G., Yang, Q., Hong, X., Wu, Y., & Li, Y. (2016). Porous molybdenum phosphide Nano-Octahedrons derived from confined phosphorization in UIO-66 for efficient hydrogen evolution. *Angewandte Chemie International Edition, 55*, 12854–12858.

Yang, J., Zhu, G., Liu, Y., Xia, J., Ji, Z., Shen, X., & Wu, S. (2016). Fe3O4-decorated Co9S8 nanoparticles in situ grown on reduced graphene oxide: A new and efficient electrocatalyst for oxygen evolution reaction. *Advanced Functional Materials, 26*, 4712–4721.

Yang, Y., Fei, H., Ruan, G., Xiang, C., & Tour, J. M. (2014). Edge-oriented MoS2 nanoporous films as flexible electrodes for hydrogen evolution reactions and supercapacitor devices. *Advanced Materials, 26*, 8163–8168.

Ye, R., del Angel-Vicente, P., Liu, Y., Arellano-Jimenez, M. J., Peng, Z., Wang, T., Li, Y., Yakobson, B. I., Wei, S.-H., Yacaman, M. J., & Tour, J. M. (2016). High-performance hydrogen evolution from MoS2(1–x)P x solid solution. *Advanced Materials, 28*, 1427–1432.

Yeo, R. S., Orehotsky, J., Visscher, W., & Srinivasan, S. (1981). Ruthenium-based mixed oxides as electrocatalysts for oxygen evolution in acid electrolytes. *Journal of the Electrochemical Society, 128*, 1900–1904.

Yu, Y., Huang, S.-Y., Li, Y., Steinmann, S. N., Yang, W., & Cao, L. (2014). Layer-dependent electrocatalysis of MoS2 for hydrogen evolution. *Nano Letters, 14*, 553–558.

Yuan, L., Yan, Z., Jiang, L., Wang, E., Wang, S., & Sun, G. (2016). Gold-iridium bifunctional electrocatalyst for oxygen reduction and oxygen evolution reactions. *Journal of Energy Chemistry, 25*, 805–810.

Zhang, C., Huang, Y., Yu, Y., Zhang, J., Zhuo, S., & Zhang, B. (2017). Sub-1.1 nm ultrathin porous CoP nanosheets with dominant reactive {200} facets: A high mass activity and efficient electrocatalyst for the hydrogen evolution reaction. *Chemical Science, 8*, 2769–2775.

Zhang, K., Kim, H.-J., Lee, J.-T., Chang, G.-W., Shi, X., Kim, W., Ma, M., Kong, K., Choi, J.-M., Song, M.-S., & Park, J. H. (2014). Unconventional pore and defect generation in molybdenum disulfide: Application in high-rate lithium-ion batteries and the hydrogen evolution reaction. *ChemSusChem, 7*, 2489–2495.

Zhang, L., Wu, H. B., Yan, Y., Wang, X., & Lou, X. W. (David). (2014). Hierarchical MoS2 microboxes constructed by nanosheets with enhanced electrochemical properties for lithium storage and water splitting. *Energy & Environmental Science, 7*, 3302–3306.

Zhang, T., Wu, M.-Y., Yan, D.-Y., Mao, J., Liu, H., Hu, W.-B., Du, X.-W., Ling, T., & Qiao, S.-Z. (2018). Engineering oxygen vacancy on NiO nanorod arrays for alkaline hydrogen evolution. *Nano Energy, 43*, 103–109.

Zhang, X., Yu, X., Zhang, L., Zhou, F., Liang, Y., & Wang, R. (2018). Molybdenum phosphide/carbon nanotube hybrids as pH-universal electrocatalysts for hydrogen evolution reaction. *Advanced Functional Materials, 28*, 1706523.

Zhang, Y., Ding, F., Deng, C., Zhen, S., Li, X., Xue, Y., Yan, Y. M., & Sun, K. (2015). Crystal plane-dependent electrocatalytic activity of Co3O4 toward oxygen evolution reaction. *Catalysis Communications, 67*, 78–82.

Zheng, T., Sang, W., He, Z., Wei, Q., Chen, B., Li, H., Cao, C., Huang, R., Yan, X., Pan, B., Zhou, S., & Zeng, J. (2017). Conductive tungsten oxide nanosheets for highly efficient hydrogen evolution. *Nano Letters, 17*, 7968–7973.

Zheng, Y., Jiao, Y., Jaroniec, M., & Qiao, S. Z. (2015). Advancing the electrochemistry of the hydrogen-evolution reaction through combining experiment and theory. *Angewandte Chemie International Edition, 54*, 52–65.

Zhong, X., Sun, Y., Chen, X., Zhuang, G., Li, X., & Wang, J.-G. (2016). Mo doping induced more active sites in Urchin-Like W18O49 nanostructure with remarkably enhanced performance for hydrogen evolution reaction. *Advanced Functional Materials, 26*, 5778–5786.

Zhou, W., Hou, D., Sang, Y., Yao, S., Zhou, J., Li, G., Li, L., Liu, H., & Chen, S. (2014). MoO2 nanobelts@nitrogen self-doped MoS2 nanosheets as effective electrocatalysts for hydrogen evolution reaction. *Journal of Materials Chemistry A, 2*, 11358–11364.

Zhou, W., Wu, X.-J., Cao, X., Huang, X., Tan, C., Tian, J., Liu, H., Wang, J., & Zhang, H. (2013). Ni3S2 nanorods/Ni foam composite electrode with low overpotential for electrocatalytic oxygen evolution. *Energy & Environmental Science, 6*, 2921–2924.

Zhou, W., Zheng, J.-L., Yue, Y.-H., & Guo, L. (2015). Highly stable rGO-wrapped Ni3S2 nanobowls: Structure fabrication and superior long-life electrochemical performance in LIBs. *Nano Energy, 11*, 428–435.

Zhou, X., Jiang, J., Ding, T., Zhang, J., Pan, B., Zuo, J., & Yang, Q. (2014). Fast colloidal synthesis of scalable Mo-rich hierarchical ultrathin MoSe2−x nanosheets for high-performance hydrogen evolution. *Nanoscale, 6*, 11046–11051.

Zhu, Y., Zhou, W., Sunarso, J., Zhong, Y., & Shao, Z. (2016). Phosphorus-doped Perovskite oxide as highly efficient water oxidation electrocatalyst in alkaline solution. *Advanced Functional Materials, 26*, 5862–5872.

Zou, X., Su, J., Silva, R., Goswami, A., Sathe, B. R., & Asefa, T. (2013). Efficient oxygen evolution reaction catalyzed by low-density Ni-doped Co3O4 nanomaterials derived from metal-embedded graphitic C3N4. *Chemical Communications, 49*, 7522–7524.

Chapter 5
Potential Applications of Electrolysis for Commercial Hydrogen Production

Abstract The fossil fuel based energy resources are considered as the primary source of energy for day today requirement. But the limited reserve and carbon emission during the combustion process restricts their use demanding an alternative resource. After numerous research efforts, the researchers have successfully stored solar energy in form of chemical energy, especially in molecular hydrogen (H_2). Like oil and natural gases, hydrogen is not energy but stores and carries energy. On the other hand, for ease of use, the online production of H_2 remain indispensable. This chapter presents the development in the use of electrolysis for commercial hydrogen production, onsite electrolysis and use of H_2 as a clean fuel in vehicles.

Keywords Fuel cell · Hydrogen · Onsite production · Electrolyzer · Centralized production

5.1 Hydrogen as an Industrial Commodity and Vehicle Fuel

Being the third most abundant element on earth, hydrogen could revolutionize the energy demand and supply, globally. It has very high specific energy per mass density, at least twice that of gasoline or similar petroleum products (IEA Hydrogen 2017). Hydrogen has long been used at industry levels for the processing/refining of crude petroleum oil and for desulfurization. Hydrogen, until recently, has largely been used as feedstock for the oil refineries and also for the production of methanol as an industrial commodity. Vast majority of these industrial processes (refining, production of methanol and ammonia, synthesis of bulk chemicals) rely on the fossil fuel based feed stocks, which tend to destabilize the climatic conditions and the living creatures thriving on them (Hydrogen Council 2017). The implementation of decarbonization in these industrial processes has been initiated and expected to escalate rapidly towards the end of 2050 to achieve the target of two-degree Celsius decrease in the global temperature (IRENA 2018). The significant component of these feed stocks is hydrogen and is derived from the natural gas through gasification, inadvertently increasing the global carbon footprint. The only option is to produce hydrogen from renewables and use it as and when required. Basically, here we speak of using hydrogen as a potential energy carrier that has an advantage over electricity. The use of clean hydrogen is on the target by numerous policy makers and industries, which is expected to gain momentum towards the end of 2035 (Global CCS Institute 2018).

The hydrogen produced on an industrial scale has been broadly categorized into three different segments, namely, brown (or grey), blue, and green hydrogen depending on the resources used and the method implicated to derive hydrogen. Steam methane reforming (SMR) is by far the most widely implemented industrial technique to produce hydrogen. This process yield both hydrogen and CO_2, which is undesired considering its large CO_2 emission signature. The release of CO_2 into the atmosphere puts the hydrogen produced via SMR technique into the grey-hydrogen category (GasTerra 2018). Figure 5.1 shows the various methods currently employed for the production of hydrogen. Now, as the world is vying for CO_2-neutrality, production of hydrogen through SMR is largely going to be replaced by a relatively new method, i.e., carbon capture and storage (CCS) (GasTerra 2018). This method takes advantage of the barren lands and empty field to store emitted CO_2 via a method similar to compressed air energy storage (CAES). CCS is a pre-carbon capture technique which filters the CO_2 during the gasification process and stores it underground for recycling or for producing other chemicals. The hydrogen produced in this case is free from CO_2. Though CO_2 is not released to the open atmosphere, it's not eliminated completely either. The hydrogen thus produced is called blue hydrogen. Thus large scale production of blue hydrogen can significantly reduce the CO_2 content in the atmosphere and can be implemented as an industrial commodity that can be used for electricity generation or automobile fuel.

Recent research and development have garnered few environmentally benign techniques that will use only electricity to split water to produce hydrogen. Methods

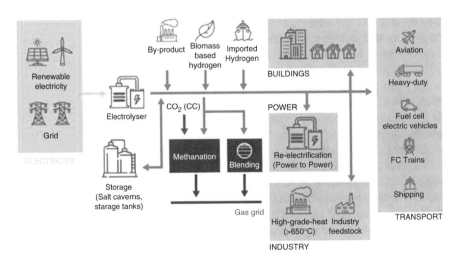

Fig. 5.1 Schematic presentation of different methods currently used for hydrogen production. (Reproduced with the permission from www.irena.org, 2018)

such as electrolysis, electrochemical, or photoelectrochemical water splitting have emerged as the next generation hydrogen production techniques with no trace of carbonization, hence ~100% decarbonization is achieved. The hydrogen produced in these techniques is called as green hydrogen as it has zero CO_2 emission (GasTerra 2018). The significant progress at industrial level to produce hydrogen as commodity has escalated recently with the depleting fossil fuel reserves and issues like global warming and greenhouse effect. In this context, hydrogen can provide a long term solution for being an industrial commodity that can be used for various purposes and can be produced on demand through smart CCS and emerging concepts such as electrolysis, electrochemical and photoelectrochemical water splitting (DOE 2014). Examples of such technologies include producing hydrogen via electrolysis using electricity (derived from renewable resources such as water, solar, wind etc.) or recycling carbon to generate syngas for synthetic fuels. These relatively new methods have been tested successfully (though at laboratory scales), and will have significant impact on the production of green hydrogen.

Aforementioned hydrogen production technologies have attained maturity over time. Scaling up the process could gradually bring down the cost. The electrolysis or electrochemical processes are expected to achieve commercial viability towards the end of 2030. Subsequent development to store and disseminate/transport the hydrogen thus produced would further improve the overall infrastructures. Technologies like proton exchange membrane (PEM) electrolyzers and fuel cells are approaching technical maturity and economies of scale. Major players in energy generation, industrial gas manufacturers, vehicle manufacturers, and other industry stakeholders of several regions of the world, e.g., the USA, Europe, and Japan, have formed Hydrogen Council to take advantage of this potentially large and rapidly growing market. They aim to make the best use of existing infrastructure (e. g. the gas grid) and to prepare for hydrogen from renewables potentially to partly replace the energy supply and revenues that are now based on oil and gas. Hydrogen can therefore play an important role in the energy transition. As stated earlier, smart CCS provides the option for both CO_2 capture and release of pure hydrogen, thus can be helpful in developing hydrogen infrastructures. This is vital for areas where post-combustion carbon capture is not possible (as it consumes significant amount of energy). It is to be noted that, as long as hydrogen is being produced from electricity generated from coal, it will be called brown hydrogen, which does not help in solving climate issues. Though blue hydrogen helps partially, green hydrogen (produced via electricity generated through sustainable techniques) remains the key to all the environmental issues.

With gradual increase in the hydrogen production, installation or up gradation of pipelines for the hydrogen transportation, storage methods, and filling stations, hydrogen would emerge as the most prolific and cleanest form of energy carrier both at domestic and industrial level. In Europe, there are at least five dedicated projects running for hydrogen storage ranging from 0.7 MW to 2 MW. Similar initiative has also been taken up by the National Renewable Energy Laboratory (NREL) in the USA. The NREL in collaboration with Xcel Energy generates electricity through photovoltaic cells and wind turbine, which is then used to electrolyze water, producing pure hydrogen. This hydrogen is then stored for future use (OilPrice.com n.d.). Though cost and efficiency could be few of the bottlenecks at initial level of hydrogen storage, we would be able to see a potential increase in the research and development of hydrogen production and storage in years to come, because of the long term benefits they have to offer: emission-free power that can also double as car fuel and raw material in a range of industrial production applications. In fact, the secretary of the Hydrogen Council, Engie's VP of advanced business and technologies Pierre-Etienne Franc, says that "The years 2020 to 2030 will be for hydrogen what the 1990s were for solar and wind" (OilPrice.com n.d.).

The first hydrogen powered automobile was built around 1991, and there have been almost 10,000 such hydrogen fuel cell powered vehicles are now running around the world ("Hydrogen economy: it's time to scale it up – Pierre-Etienne Franc" n.d.). The move from fossil fuel based internal combustion engines to clean-energy (zero-emission) based electricity or hydrogen fuel cell powered internal combustion engines have gone rapidly during the past few years. It is estimated that hydrogen has roughly twice the amount of specific energy than gasoline. Plus, the steadfast progress to achieve the green and blue (at least) hydrogen production scale to commercial viability could see, in near future, a revolutionary change in both the light and heavy transportation modes, worldwide. According to an estimation by the "Hydrogen Council", the number of hydrogen fueling stations worldwide should increase from 375 in 2017 to more than 1100 in 2020, everywhere from the state of California to Japan and South Korea, Denmark and Germany (Hydrogen Council 2017). In this context, China has plans to have roughly 3000 hydrogen-powered buses operating in Shanghai alone, with Paris, the capital of France, not lagging behind far from the rest (Hydrogen Council 2017). Even though the demand and supply of hydrogen based power is still at a nascent stage, we can see that the current production capacity can easily power a staggering 10 million hydrogen-powered vehicles. Meanwhile, major players like Air Liquide is developing its Blue Hydrogen initiative, and has strong commitment to produce at least 50% of the hydrogen dedicated to energy applications through carbon-free processes by 2020 (Hydrogen Council 2017). During the Global Climate Action Summit, active members of Hydrogen Council, i.e., Air Liquide and Hyundai, stayed firm to their goal of achieving 100% decarbonized hydrogen fuel in the transportation sectors by the year 2030. However, the current approach towards clean hydrogen energy policy should further be accelerated and parallel development regarding the support infrastructures should also be promoted. Hydrogen can be used, besides as a vehicular fuel, in feed stocks, for heating and generating power for household, power generation and storage, which will significantly push the process of large scale and rapid decarbonization in key industrial and other heavy transportation sectors. In transportation, hydrogen-powered vehicles are commercially available now or will become available in the next 5 years in medium-sized and large cars, buses, trucks, vans, trains, and forklifts (Dodds et al. 2015). In these segments, FCEVs meet the performance and convenience requirements best. In the next wave, costs are likely to drop with scale, allowing hydrogen to compete in more segments such as smaller cars and minibuses. By 2030, 1 in 12 cars sold in California, Germany, Japan, and

South Korea could be powered by hydrogen, more than 350,000 hydrogen trucks could be transporting goods, and thousands of trains and passenger ships could be transporting people without carbon and local emissions ("Hydrogen economy: it's time to scale it up – Pierre-Etienne Franc" n.d.). Beyond 2030, hydrogen will increasingly be used to create renewable synthetic fuels to decarbonize commercial aviation and freight shipping, which are harder to decarbonize using pure hydrogen and fuel cells. Other clean modes, like battery electric systems, or hybrid smart vehicular systems rely on electricity to run. Though these have advantage over hydrogen fuel cell run vehicles, on the long run, hydrogen could readily complement or more likely replace the mode of electricity generation to boost the automobile industries. However, such achievements would require efforts from all the sectors, be it public, private, or academic, and the approach needs to be collective worldwide to make the hydrogen revolution, a success story.

5.2   Growing Markets for Pure Hydrogen Products

The global demand for pure hydrogen based fuel/products could reach as large as ten-fold, from 8 EJ in 2015 to almost 80 EJ in 2050 (see Fig. 5.1 courtesy of the Hydrogen Council) (Hydrogen Council 2017). Hyundai Motor and Yield Capital plan to raise $100 million for a Hydrogen Energy Fund, predicting that numerous new hydrogen startups will come about over the coming years. Hydrogen production can be pivotal for both household and industrial reforms in the context of renewable energy. Considering the gradual pace at which innovation and adoption of hydrogen has been occurring, the global market would see an explosion of pure hydrogen based products as late as 2030. Ambitious yet timely collaborations between government sectors and industries have been deployed to incur cutting edge technologies to address the challenges currently faced by the hydrogen storage systems, and to reduce the overall cost of production. These actions enabled new hydrogen applications to expand beyond vehicles, with hydrogen technology now powering villages, enabling hybrid storage systems for renewables, and moving into the aerospace and marine sectors. Figure 5.2 shows the market trends towards the growing demand for hydrogen products on a time-scale.

According to the Cleantech Group, "There are over 60 million tons of hydrogen produced annually, worth almost 100 billion USD (Cleantech Group 2018). At present, almost 80% of the hydrogen generated is being used for three main industrial applications: refineries, ammonia production or metal processing. The latest "Economist Technology Quarterly" also points to the growing influence of hydrogen, calling it "the most promising technological solution to decarbonize our energy system" (Cleantech Group 2018). Another issue that grapples the hydrogen economy is its distribution. This can be solved, to an extent, by utilizing the existing gas lines and infrastructural supports. That could see a green way of heating households/industries and generation of electricity throughout the ear even when the weather conditions are neither sun-rich, nor windy. Research and development in this context have already

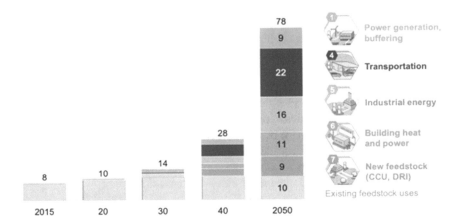

Fig. 5.2 The market trends towards the growing demand for hydrogen products on a time-scale. (Reproduced with permission from Hydrogen Council 2017)

been progressed at a global level, and scalability in hydrogen production to cope with the energy demand is expected to be achieved around 2030, with the equivalent of more than 5 million households connected to a gas network with blended or pure hydrogen. Once the hydrogen economy stabilizes, in the next step, cost-sensitive industry level assignments would be optimized with a targeted time-scale. It is expected that the penetration of hydrogen as an energy carrier would be not that promising in the industrial processes, at initial stages, and may require collective support from other renewables. However, with large research base and substantial funding, this issue could be resolved with time. By 2030, up to 200 steel, chemical, and automotive plants could be pioneering the use of hydrogen for heat and power. All the energy sources that have been enlisted to facilitate hydrogen production are mostly renewables. Hydrogen can be used in multifold ways, i.e., it can store electricity from the renewable resources in itself for indefinite period of time, and can produce the same in a carbon-clean way. Hydrogen is thus a key enabler of the transition to renewable energy. By 2030, 250 to 300 TWh of surplus renewable electricity could be stored in the form of hydrogen for use in various other segments. In addition, more than 200 TWh could be generated from hydrogen in large power plants to accompany the transition to a renewable electricity system (IEA Hydrogen 2017).

5.3 Electrolysis at Onsite Use and Centralized Production Scales

Though hydrogen can be a potential energy carrier that could solve myriads of energy issues and crisis all over the world, it suffers from one significant challenge, i.e. storage. Storage of hydrogen requires large spaces and that too at a recommended pressure of 50 atm. One solution is to liquefy the gaseous hydrogen and distribute just like the household gas cylinders. However, compressing hydrogen requires high pressure and is a costly affair. Also, low temperature is highly essential in keeping the stored hydrogen stable. Hydrogen has several characteristics that limit its storage and transportation. It has the lowest ignition threshold, i.e., minimum of energy required to burn hydrogen in comparison to any other fuels. In addition, it has the widest flammability range of all the gases and it can readily escape through even the tiniest hole possible (due to its extremely small size) (Morgan 2006).

As hydrogen is classified as a highly hazardous material, the separation between its storage and other site operations is strictly regulated according to the amount of hydrogen being stored. Also, hydrogen storage requires quite a large space due to its low specific energy to volume density. In the United States, the maximum limit for indoor hydrogen storage is generally 2999 Scf – about 10 cylinders. This amount easily scales up when it comes to industrial grade applications of hydrogen, where storage of hydrogen at large scale becomes a huge concern ("Is On-Site Electrolysis Hydrogen Generation a Fit for Your Hydrogen Requirement?" n.d.). Thus, for industrial applications, on-site hydrogen production is essential which would generate hydrogen as and when required, preventing the need of high volumetric storage systems (which are costly). Polymer electrolyte membrane based water electrolyzers provide cost-effectiveness in operational, production and safety that use very little hydrogen to be served by hydrocarbon reforming. Previously, these users would have had to utilize delivered and stored hydrogen. The advantage with PEM on-site hydrogen generators is that the cost associated with the overall process including small-to-medium flow rates of pressurized, highly pure hydrogen can be easily predicted, and doesn't require costly storage techniques (near-zero hydrogen inventory). These generators can vary production rates to suit the requirements of customers.

PEM based on-site hydrogen production offers the following advantages over stored/deliverable hydrogen;

1. It requires near-zero gas inventory
2. The overall equipment compactness and safety features don't require tedious safety review processes over the time or additional storage/emission permissions.
3. On-site electrolyzer produces water-free and high gas purity hydrogen
4. Load following by varying the rate of hydrogen delivery exactly as required
5. Fast permitting, easy installation, simple operation, high reliability, and minimal maintenance
6. Medium gas pressure: high enough pressure to allow for surge storage load leveling

Centralized hydrogen production is thus ideal to bring economies of scale, where large quantity of hydrogen can be produced and later can be distributed in bulk amounts through gas-line infrastructures. In contrast, smaller energy requirements would not favor centralized production. They would rather opt for distribution or

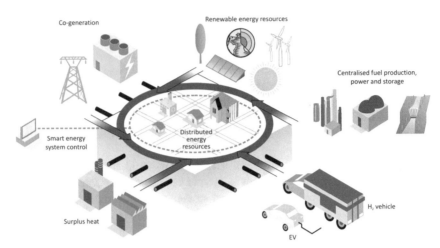

Note: EV = electric vehicle; H₂ = hydrogen.
Source: IEA (2013a), *Transition to Sustainable Buildings: Transition and Opportunities to 2050*, OECD/IEA, Paris.

Fig. 5.3 Schematic presentation of energy system as an intelligent energy network. Note: EV electric vehicle, H_2 hydrogen. (Reproduced with permission from Eisentraut, Anselm, and Adam Brown, www.iea.org, 2013–14)

localized hydrogen production. This would provide hydrogen on demand, however, moving the price index higher. Localized hydrogen production is ideal for those who have temporary, unpredictable hydrogen demand and does not guarantee high purity. Figure 5.3 shows provides an illustration of centralized and distributed hydrogen energy system (Eisentraut and Brown 2014).

Thus, the most affordable solution can be in the form of onsite production of hydrogen and that too through centralized plants, which would provide bulk hydrogen to meet the ever growing energy demand. This can only be achieved through electrolysis process. The yielding mass being the green hydrogen, because electrolysis makes use of water to produce hydrogen (green hydrogen to be specific) without affecting the environment.

References

Cleantech Group. (2018). The role of green hydrogen in global decarbonization.
Dodds, P. E., Staffell, I., Hawkes, A. D., Li, F., Grünewald, P., McDowall, W., & Ekins, P. (2015). Hydrogen and fuel cell technologies for heating: A review. *International Journal of Hydrogen Energy, 40*, 2065–2083.
DOE. (2014). Alternative Fuels Data Center: Hydrogen production and distribution [WWW Document].
Eisentraut, A., & Brown, A. (2014). Heating without global warming - market developments and policy considerations for renewable heat, Iea.
GasTerra. (2018). Hydrogen and CCS: A smart combination.

Global CCS Institute. (2018). Global Carbon Capture and Storage Institute Submission on the Development of a National Hydrogen Strategy Introduction to the Global Carbon Capture and Storage Institute What do you think are the two or three most significant recent developments in hydrog.

Hydrogen Council. (2017). Hydrogen scaling up : A sustainable pathway for the global energy transition. www.hydrogencouncil.com. Hydrog. Counc. Hydrog. scaling up a sustain. Pathw. Glob. energy transition. www.hydrogencouncil.com 80.

Hydrogen economy: it's time to scale it up - Pierre-Etienne Franc, (n.d.).

IEA Hydrogen. (2017). Global trends and outlook for hydrogen, prepared by mary-rose de Valladares.

Is On-Site Electrolysis Hydrogen Generation a Fit for Your Hydrogen Requirement?, (n.d.).

Morgan, T. (2006). The hydrogen economy. A non-technical review. https://wedocs.unep.org/handle/20.500.11822/9024.

OilPrice.com. (n.d.). Will hydrogen break the battery market?

IRENA. (2018). Hydrogen From Renewable Power: Technology outlook for the energy transition.

Chapter 6
Summary and Conclusion

Abstract This chapter summarizes the ongoing development in electroactive material design and electrode fabrication techniques. Also the scope to work to further improve the efficiency of the electrolyzer is presented.

Keywords Gaseous fuels · Noble metals · Metal oxide/hydroxide · 2D · 3D · Metal foils

The electrochemical splitting of water has been considered as one of the efficient process for the scalable evolution of gaseous fuels like hydrogen and oxygen. Both the processes are proceeds through a multi proton/electron path ways, thereby showing higher reaction activation barrier experimentally. Therefore various electrocatalysts were developed and employed to reduce this potential barrier. Furthermore on the basis of both the theoretical and experimental works, the electrocatalytic performance of catalysts strictly depends on the extent of exposed active sites. Hence the designing of nanostructured materials finds a promising way to towards excellent catalytic activities in terms of selectivity, specificity and durability of a particular electrochemical reaction. As discussed in the above section as well as in the previous reports, the noble metal (like Pt, Ru, Ir, Rh etc.) based electrocatalysts shows efficient catalytic performances for both the HER and OER. But the high cost and low abundance restricts these noble metal electrocatalysts from the commercial application. Moreover the development of low cost materials with earth abundant elements through a facile synthetic method is indispensible for wide spread application. In this regard many of the scientific efforts have been devoted and different non-noble metal based low precious electrocatalysts have been developed. Among them the transition metal chalcogenides, metal oxides/hydroxides, nitrides, phosphides were shows profound catalytic activity. But their performance still remained far apart from the activity of the noble metal based catalysts demanding further modification and development. Afterwards the composites with conductive carbons like graphene, hetero atom doped graphene, and carbon nanotubes etc. are designed. All the electrocatalysts thus developed are in powdery form and demonstrates lower catalytic durability, which may be due to the weaker contact of the catalyst with the electrode causing peeling out of the catalyst layer from the electrode. In order to resolve this problem the self-supported and free standing electrodes were designed by following various synthetic protocols. Generally supports like 2D metal foils (nickel foil, copper foil, aluminum foil, gold foils etc.), 3D metal foams (nickel foam, copper foam, stain less steel meshes etc.), carbon cloths, carbon fiber, carbon foams etc. are developed and employed for the preparation of free standing and self-supported electrodes.

Although much more materials have been developed in last few decades, but still more research is needed to evolve new electrocatalysts to meet the catalytic performances of the noble metal based materials. (a) Better characterization tools like in situ Raman and X-ray absorption spectral analysis should be employed to find out the actual active reaction sites experimentally and to gather more information regarding the reaction pathways followed during the HER and OER by the non-noble metal based electrocatalysts. (b) Basing upon the theoretical calculations and experimental results, some sensible reaction models close to the reaction systems should be developed to detect the actual reaction intermediates and OER/HER mechanisms followed on these electrocatalyst surfaces. These fundamental insights not only assist to design new electrocatalysts but also will be helpful for their performance enhancement and to establish the structure reactivity relationship among them. (c) Further the catalytic activity of a particular electrocatalyst (may be TMC, TMO, TMN, TMP etc.) towards OER/HER in a specific electrolyte (either in acidic, neutral or alkaline) should be optimized for practical application. (d) More focus should be paid on designing of the set up configuration to integrate these HER/OER catalysts into other renewable power sources like solar energy, wind energy, metal air batteries etc.

Index

© The Author(s), under exclusive license to Springer Nature Switzerland AG 2019
A. K. Samantara, S. Ratha, *Metal Oxides/Chalcogenides and Composites*,
SpringerBriefs in Materials, https://doi.org/10.1007/978-3-030-24861-1

Printed in the United States
By Bookmasters